Introduction to

SCHOLARSHIP

Building Academic Skills
for Tertiary Study

CHERYL SIEWIERSKI

OXFORD
UNIVERSITY PRESS
SOUTH AFRICA

Oxford University Press is a department of the University of Oxford.
It furthers the University's objective of excellence in research, scholarship,
and education by publishing worldwide. Oxford is a registered trade mark of
Oxford University Press in the UK and in certain other countries.

Published in South Africa by
Oxford University Press Southern Africa (Pty) Limited

Vasco Boulevard, Goodwood, N1 City, Cape Town, South Africa, 7460
P O Box 12119, N1 City, Cape Town, South Africa, 7463

Introduction to scholarship: Building academic skills for tertiary study 2e

ISBN 978 0 19 074146 4

First impression 2020

Typeset in Utopia Std 9.5pt on 12pt
Printed on 70gsm Woodfree Bond

Acknowledgements
Publishing Manager: Alida Terblanche
Publisher: Marisa Montemarano
Development Editor: Jeanne Maclay-Mayers
Editor: Megan Hall
Indexer: Mary Lennox
Typesetter: Lumina Datamatics, Ltd
Art Director: Judith Cross
Spec Designer: Judith Cross
Cover Designer: Judith Cross
Cover Image: Aleksandr Andrushkiv, Shutterstock; Sharna Sammy, OUP
Printed and bound by: ABC Press
14458

The authors and publisher gratefully acknowledge permission to reproduce copyright material
in this book. Every effort has been made to trace copyright holders, but if any copyright
infringements have been made, the publisher would be grateful for information that
would enable any omissions or errors to be corrected in subsequent impressions.

Links to third party websites are provided by Oxford in good faith and for information only.
Oxford disclaims any responsibility for the materials contained in any third party website
referenced in this work.

CONTENTS

ABOUT THE AUTHOR

Cheryl Siewierski is a Princeton-based South African academic who has been in the education arena for more than 20 years. She began her career teaching high school English and Criminology, and then went on to lecture for both Unisa and The Independent Institute of Education (The IIE) at Varsity College, where she received awards for Teaching Excellence, and for Innovations in Teaching and Learning.

Since then, she has worked both permanently and on contract for The IIE. As Senior Head of Programme and acting Head of Faculty Social Sciences, she managed The IIE's Academic Literacy programme, its Postgraduate Diploma in Education, and various English language and literature modules. She designs and develops curriculum material and instructional design modules for English literature, Criminology, Media Studies, and Psychology, and also lectures Ethics and Philosophy of Education online. Her approach in all of these endeavours is designed to help students develop into critical, adaptable thinkers who are ready for ever-more nebulous academic and working environments. Cheryl is an external moderator and referee for other tertiary institutions and journals, and co-author of a popular South African business communication textbook.

Cheryl obtained a Master's degree in Education from the University of the Witwatersrand with distinction in early 2015. She holds an Honours degree in Criminology from the University of Pretoria, a Bachelor of Arts degree with triple majors in English, Psychology and Criminology, and a Higher Education Diploma from the University of Pretoria. Committed to lifelong learning, Cheryl has also studied various environmental science and management subjects at Unisa as an occasional student, and is currently completing her PhD in Higher Education Policy and Leadership at the University of Pretoria.

When she is not working, Cheryl enjoys reading and conversing about current events, ethics, economics, sustainability, nature, and technological innovation, and relaxes by travelling and spending time with her husband and dogs in Princeton, New Jersey.

FOREWORD

One of the biggest mistakes we make about students entering university is the assumption that they can do basic academic tasks reasonably well. We assume high school students can read. We assume they can write. We assume they can make a public argument. We assume they can evaluate rival positions and make their own stand on a difficult subject. We even assume they know something about what's happening in the world around them from economic growth rates to youth unemployment.

We are then startled as professors that undergraduate students fail in large numbers; that they find difficulty writing essays in which the logic from one paragraph flows seamlessly into another; that their arguments are drowned in emotion when the subject matter becomes personal or controversial; that their social knowledge is weak; and that public argument often degrades into personal insult.

It is not difficult, of course, to understand how South African youth came into this sad state of affairs. We do not teach students to learn deeply but rather to pass tests. Few students read a complete book thoughtfully and critically except, in the case of a prescribed work, for examination. Writing long essays is a thing of the past with multiple choice questions preferred for ease of assessment. Teachers seldom show up every single day to teach in the majority of our schools, and few are interested in linking the subject matter to the world outside the classroom. The resultant tragedy is as predictable as a drought in the Karoo – severe under-preparation for university studies.

Then along comes a book called *Introduction to Scholarship* that empowers students to make the link between school and university, between disciplinary knowledge and strategic knowledge (how to make it through university), and between knowing and doing academic work. This excellent resource, which every university student must possess, makes no assumptions; it understands that there is a gap between what high school students know and what is required to successfully navigate the academic journey towards a degree.

All bases are covered from planning and writing an assignment, to using search engines efficiently, to reading with purpose, to evaluating evidence, and to the role of rhetoric in argument. This is the ultimate 'bridging module' that prepares students for reading, writing, thinking and analysing in university environments. Objectives are clearly composed; hands-on tasks and lively examples appear throughout the text; and there are helpful references to supporting sources. And unlike so many other texts on the subjects, the content is accessible and even fun – explore, for example, the section on Black-belt 'Google-Fu'.

It will take a long time for the South African school system to be overhauled in order to serve the needs of all 13 million children in public and private schooling. It is unlikely that every school would have an outstanding academic guidance counsellor that prepares school leavers for further study. It might never happen that schools produce 'teachers as intellectuals', an insertion in the literature on critical theories of schooling when we all still had hope that the post-apartheid period would usher in not simply more education but better education.

Until then, we are dependent on high quality resources that enable students to migrate as seamlessly as possible from school to university, and from university to work. I believe that one reason that students struggle at university is ignorance – they simply do not know. Few students understand, for example, how to do basic referencing of other people's work and that is why plagiarism is so rife on campuses. Nor have students grappled with basic human skills such as what it means to be ethical, so that even those who went to schools where they were taught about plagiarism find themselves trapped when directly using sources without acknowledgment.

But ignorance and ethics explain only part of the problem of academic preparation; it is also habits of mind. Some researchers speak of successful students as having adopted 'an academic identity' where certain qualities adopted enable them to succeed in higher education. This new book can only do so much in terms of providing critical information – it must still be picked up, read and its key ideas made part of that academic commitment of the student. In other words, this invaluable book in the hands of a facilitator working with a group of undergraduates can have an even greater impact on the academic skills required by university students.

Nor should student growth stop with the acquisition of these foundational skills in what could be called, broadly speaking, academic literacy. We have a duty as academic teachers to launch students into much deeper learning at school but especially in university. I really believe that no student should graduate without reading Machiavelli's *The Prince* or Hawking's *A Brief History of Time* or Fanon's *Black Skin, White Masks* or Mda's *Madonna of Excelsior*. This is called *education* and must be distinguished from training, the acquisition of technical knowledge and skills in Quantity Surveying or Physiotherapy or Teacher Education or Private Law. Learning to read or write at the level required by universities is in and of itself a basic occupation; applying those insights to what we call *higher* education is in fact what this enterprise should be about. This book would have failed, therefore, if it did not catapult the user into these higher trains of thought.

Professor Jonathan Jansen

Distinguished Professor in the Faculty of Education, University of Stellenbosch

PREFACE

Having done the 'smart' thing by first taking a gap-year and only then launching myself at university as a 19-year-old, I thought that I had my university life figured out: I had some practical work experience, I knew what I wanted to study (well, sort of), and was ready to 'get stuck in' to student life.

Of course, I was stupendously unprepared for tertiary studies.

It didn't matter that I had done well in school, or that I was reasonably smart. When my entire class was failed for our first English assignment because we hadn't referenced correctly, or at all (referencing? What *is* that?), I realised that I was dealing with an entirely new world: one that I had no travel guide for. Lecturers would award low marks for work I felt was brilliant, without identifying particular problems or indicating ways in which I could correct them, and it was clear that my, and my classmates', school-assignment-approach was horribly inadequate for university. I watched as too many of my really smart classmates dropped out of university in that.

Taking a determined 'Eye of the Tiger' view of my studies though, I refused to become part of the first-year attrition statistics, and I worked consciously and conscientiously to figure things out, step-by-awkward-step, with my marks ultimately improving each year. Not having the benefit of 'Professor' Google in those early years of study, my learning was more tortoise, and less hare in terms of speed! Ultimately, while the pace of my learning has increased exponentially since technology came to the rescue, my learning about learning, and how to get the most out of studies, continues to this day.

This book is in essence then, a guided tour through important things about learning at university that I learnt the long, hard way, with this edition featuring an entirely new chapter on digital citizenship, additional content on active listening, frames of reference, and a basic primer on research paradigms. It also features updated and additional content throughout. Once again, through the smart and quirky, hipster tour guide of Professor Bloom, the book aims to provide you with the travel guide that I, and my fellow students so many years ago, would have found invaluable in our studies.

I hope that your regular application of this collection's (occasional) theoretical and (mostly) practical advice alleviates at least a little of the anxiety you might be feeling, and that it helps you to adapt to, and come to love, this admittedly intimidating, but wonderful, liberating, and challenging part of your life: your university years!

Warm regards,

Cheryl Leigh-Anne Siewierski

HOW TO USE THIS BOOK BEST

Students, if you're wondering who the 'hipster dude' from the book's cover is, it's time to introduce you to Professor Bloom, our little reinvention of a great educational psychologist who developed ideas on mastery learning – something we hope that you will develop with the help of his guidance. You will therefore notice several Prof Bloom icons as you work through this book – they signal various activities or resources. We have listed the icons and their function below. Make sure you know what each one means, and use the extra material (especially the activities!) that they provide.

 This icon indicates that there is relevant Learning Zone material available – this could be an online activity or assessment, an activity, or a resource, such as a fuller version of something in this book. You can find the Learning Zone here: https://learningzone.oxford.co.za/ A clear link to the exact Learning Zone item is usually given, and we've made it clear whether the resource is for students or lecturers.

 This icon indicates that recommended reading is available for those of you who wish to study a particular topic further.

 This icon indicates a point, question or activity that prompts you to reflect on what you have just learnt.

But wait, there's more!

A smart series of videos covering the cornerstones of academic practice is available here: https://www.oxford.co.za/page/98-educational-solutions-bloom-tv/ The topics include: Bloom's Taxonomy, assignment writing, referencing and note taking. The series is part of a campaign to help students learn more effectively and equip lecturers to teach better, created by Oxford University Press South Africa in partnership with the Stellenbosch University Language Centre and A Blind Spot Productions.

Professor Bloom features on the videos and also makes appearances throughout this book as in the icon on the right.

CHAPTER 1

Engaging in academic study

I am driven by two main philosophies: know more today about the world than I knew yesterday…[a]nd lessen the suffering of others. You'd be surprised how far that gets you.'
– Neil deGrasse Tyson (2012)

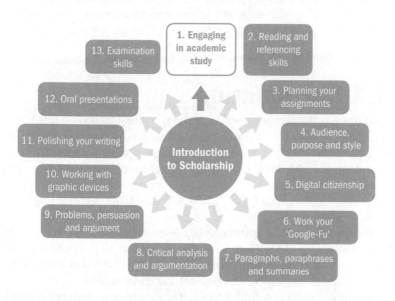

1. Engaging in academic study

2. Reading and referencing skills

13. Examination skills

3. Planning your assignments

12. Oral presentations

Introduction to Scholarship

11. Polishing your writing

4. Audience, purpose and style

10. Working with graphic devices

5. Digital citizenship

9. Problems, persuasion and argument

6. Work your 'Google-Fu'

8. Critical analysis and argumentation

7. Paragraphs, paraphrases and summaries

OBJECTIVES

At the end of this chapter, you should:

- show awareness of the differences between secondary and tertiary education;
- recognise appropriate values and attitudes for successful engagement in academic tertiary study;
- be aware of different thinking and learning styles;
- identify different cognitive strategies;
- understand the importance of reflecting on academic experiences.

IRL

Sureshni and Joseph are first-year students at the University of Kwa-Zulu Natal. Both are excited to begin their studies, but not for the same reasons. Sureshni, who has spent her high-school career studying hard and working diligently, easily managed to gain entry to the BCom programme, and is determined to work towards a well-paid career in accountancy. Joseph, however, is not too sure what he wants for the future but has, nevertheless, been admitted to the BSc Engineering programme. Although he thoroughly enjoys understanding how things work, and performed quite well at school, he is really more excited about staying in 'res' – away from the watchful eyes of his overprotective parents – and in socialising, than in acing his degree.

1.1 Introduction

Diligence refers to steady and careful effort with something.

The term **cum laude** is Latin and means 'with praise'. It is usually applied to diplomas or degrees to indicate work of distinction.

Your **identity** refers to the relatively stable ways that you define yourself in terms of your different roles (like language, religion, culture etc.). In the case of **academic identity**, it refers to how academically capable you see yourself – studies argue that the more you see yourself as 'academic', the better you will do (obviously, you need to work too!).

Independent learners are people who are able to figure out what information they need to answer a question or resolve a problem; find relevant, credible information; understand it; apply it to the problem, and then report on it.

What are your reasons for studying? Have you thought about this?

In the IRL scenario above, which of the two students do you think has a greater chance of academic success? Sureshni may seem like the obvious choice because she has a clear purpose and has demonstrated past diligence with her studies. Joseph, given his plans to party through his degree, might appear to be headed for academic disaster. But is he really doomed to drop out, and is Sureshni really a sure thing for a *cum laude* degree?

The answer to these questions is, quite simply, no – excellent high school students like Sureshni can fail – sometimes spectacularly – in tertiary programmes, and more 'average' high school students like Joseph *can*, and often do, emerge with excellent results.

Although this may sound strange, it is important to understand that university success is not *reliably* determined by past performance. Instead, studies show that the adoption of an 'academic identity' (Vincent and Idahosa, 2014; Was et al., 2009) and being an independent learner are much more important success factors at university. We will refer to both of these concepts regularly throughout the text, and will try to develop them as you progress, so do not worry if you are not yet an independent learner, or if you do not yet have a healthy academic identity.

As you settle into your studies over the next few months, you will undoubtedly notice some differences in the requirements of school and tertiary studies. While these differences are exciting, they can also be intimidating and difficult to navigate without guidance. Tertiary study demands independent work and a lot more critical thinking than study at high school does, so you might feel as though you have missed some really important skills between school and your new environment. Don't panic though – the feeling of being ill-prepared at the 'tertiary-jump' is quite normal for most students.

To help you ensure that your own jump ends in a safe landing, we would like to welcome you to this textbook, which will help you to orient yourself towards new ways of thinking about information and ideas. The remaining chapters of this textbook will guide you through a variety of topics related to thinking and learning, including:

- *Reading and referencing skills*: okay, different types of reading and reading strategies might not sound exciting, and Harvard referencing doesn't sound much better, but we can almost guarantee that you'll get a kick out of mastering those references, and you'll boost your results if you use reading strategies appropriately;
- *Planning your assignments*: this is a chunky, but really useful chapter that deals with everything you need to know about planning your writing assignments, from analysing your question to editing your work;
- *Audience, purpose and style*: this chapter looks at pitching to your reader and ensuring that you remain focused, formal and objective where necessary;
- *Digital citizenship*: this new chapter highlights key cyberethics issues and then gives helpful advice on keeping safe and staying ethical while finding and sharing information on social media and other online environments;
- *Work your 'Google-Fu'*: coaches you on how to source, collect and evaluate information using search engines and academic databases;
- *Paragraphs, paraphrases and summaries*: from developing main ideas, topic sentences, and supporting evidence, to linking them meaningfully, this chapter provides you with simple step-by-step suggestions for writing summarised and paraphrased notes for academic purposes;
- *Critical analysis and argumentation*: in this chapter, you will have fun identifying and evaluating claims and different forms of reasoning and argumentation, distinguishing between fact and opinion, and recognising different logical fallacies (so that you can avoid them!). New to the chapter are some basics on ontology, epistemology, and key academic paradigms (and in case these all sound ridiculously unpronounceable, we promise to keep it simple so you will be chewing through the terminology in no time);
- *Problems, persuasion and argument*: includes useful approaches to problem-solving, and provides theoretical and practical methods to build sound arguments – you will need to practise these throughout this book's activities and in your tertiary studies as a whole;
- *Working with graphic devices*: guides you through interpreting graphic devices, detecting and avoiding statistical bias in graphics, and in using popular software to create graphic devices;
- *Polishing your writing*: okay, we know you don't want to, but we really do need you to review parts of speech, sentence structure, paragraph-writing and punctuation so that you can submit your best assignment. Don't worry though: we'll make this as painless as possible!
- *Oral presentations*: this very useful chapter guides you through some excellent presentation skills and strategies. It will help you

to prepare speeches and oral reports, and to design presentation slides using popular software;

- *Examination skills*: helps you with creating practical study schedules, preparing for and writing your examinations, and dealing with stress and anxiety. It provides you with strategies for active listening and dealing with specific types of questions you might face too.

This first chapter will direct you to some of the differences between basic education and tertiary-level studies, and will familiarise you with the values, attitudes and skills required in post-school studies, providing you with a snapshot of the tertiary environment. It will introduce you to different learning styles, some basic approaches to thinking, and to the importance of reflecting on thinking and learning. These skills should be incorporated and practised throughout your studies, and will assist you in building a solid academic identity of your own.

Please remember to *use this textbook with all of your modules*, and not as a stand-alone text. Its design is practically focused so trying out the strategies in your personal curriculum is the best way to make sure that you benefit fully from the material. It is also important to combine the theory from this book with the practical activities available in the online Learning Zone, and with any student portal system activities provided by your academic institution.

Finally, you may already have noticed Word Use Notes in the margins of the book. These little boxes explain words that you may be uncertain of. Please pay attention to these and any other margin notes.

1.1.1 Making the leap from school to university

To be **liberated** means to be free from restrictions or conventions, either social or legal. For example, 'She found it liberating to be living on her own away from her parents.'

To **remonstrate** with someone means to express disapproval of what they are doing. For example, 'Angie was tired of the constant remonstrations from her parents about her poor results in Mathematics – she was doing the best she could.'

In southern Africa, '**res**' is an abbreviation for a hostel residence of an educational institution. For example, 'Ayanda loved living in res, but he was getting really tired of having to cook for the seniors.'

Tertiary studies, simply put, are vastly different from school studies, especially in terms of new-found freedoms. Although this can be liberating, it can also be frightening. The old cliché of the big fish in a little pond at school becoming a little fish in a big pond at university may seem like an understatement, especially at the beginning. It may feel more like being dropped in an ocean in the middle of *Shark Week*, with no map or plan to get to safety.

For example, you will not have a teacher begging you to complete your projects – if you fail to submit an assignment, you will be awarded zero. Simple. No remonstrations from your lecturer.

You might also experience the new-found freedom of living in 'res', or away from your parents for the first time. If this is the case, you may suddenly be able to decide when, or if, you are going to study. While this may feel wonderful at first, the freedom to decide not to study may end up affecting your studies quite negatively, especially given how often you can socialise in a tertiary residence.

Similarly, class attendance is usually voluntary – if you miss class, you simply miss out. There will be no formal criticism for 'bunking', as was the case at school. Sounds great, right?

The downside to this is that lecturers will also *not* be waiting for your return, eager to give you the notes from the lectures you missed. As **Voltaire** (and Spiderman, too) would remind you, it is important to remember that 'with great freedom comes great responsibility'.

Critically, you also will not succeed with simple 'parrot-fashion studying' at the tertiary level. Being able to recall lists of things is quite useful, but the volume and nature of work at tertiary level means that this is not enough. Complete the following short reflection activity to see what we mean by this.

Biographical note

Voltaire was a French Enlightenment-period philosopher who published many works advocating freedom of expression, freedom of religion, and the separation of church and state.

1. Spend one minute studying the following concepts, expressed here in Chinese symbols. You will need to be able to match up each of the concepts after studying them, so pay attention to their relationships:

   Concept A   Concept B
   七   号
   儿   人
   女   个

2. Now, cover up these concepts, and match the appropriate Concept As to the appropriate Concept B from the given options:

Concept A	Concept B
1. 儿	a) 个
	b) 人
	c) 号
2. 女	a) 个
	b) 人
	c) 号

3. Check your answers against the concepts provided under point 1 above. Did you match the related concepts correctly? Because there were only three, and you studied them for a minute, you may well have matched them correctly.

But the question is: *what have you learnt from doing so?* Do you have any ideas about what the concepts mean? Did you suddenly develop an ability to understand Chinese symbols? Do you understand how the pairs are related to each other? Are you even sure that they *are* related to each other? If I asked you to match a different set of Chinese symbols with Concept A, would you be able to respond appropriately?

The point is to realise that the benefits of memorising and repeating certain information are limited, although you might be able to do so. If you do not understand the content and the relationships between ideas, it is almost impossible to engage in more advanced cognitive activities, which are vital for academic studies.

This is good news for those who struggled with 'parrot-fashion' learning at school and preferred to understand how things work, but it does mean that constant active engagement is necessary. Keep this in mind, and remember to connect as many dots as you can by working to understand ideas, rather than simply trying to remember them.

You therefore need to train yourself to engage critically with your curriculum content, and not just study it so that you can recall it.

Table 1.1 summarises some of the main differences between school and tertiary environments that you might not yet be aware of.

Table 1.1. Some of the differences between secondary and tertiary environments

Characteristic	Secondary environment	Tertiary environment
Physical environment	Schools are usually smaller than tertiary learning environments, with two or three buildings. They are easy to find your way around.	Tertiary institutions are usually large, with multiple buildings, residences, and laboratories. They can also have multiple campuses and it may be difficult to find your way around.
Attendance	Compulsory. Any absence requires a note from your parents and is followed up.	Not usually compulsory, except for tutorials (compulsory, scheduled, small-group sessions). You may usually come and go around campus as you please.
Time management	Time is managed and arranged for you and most of your learning happens in the classroom. There is more class-time on each subject than in university, taking into account the amount of work you need to cover.	You manage your own time. Independent work is required – usually 2 to 3 hours outside of lectures for each 1 hour of class time. There is much less class-time for each subject, taking into account the amount of work you need to cover.
Assignments	Usually short assignments that are discussed and dealt with in detail in class. Teachers remind you about due dates and chase up any work not submitted.	Assignments require a great deal of reading and research, and may never be outlined in class. It is your responsibility to check due dates and submit – lecturers do not follow up work not submitted.
Test and exam scheduling	Teachers try to arrange tests and exams so that they do not conflict with each other or with other school events.	Tests and examinations are often scheduled without consideration of other courses or events – you may need to manage two or more assessments on one day!
Teaching styles	Usually face-to-face, with most learning occurring in the classroom. Teachers work from the textbook, give you notes, write down important concepts, and give you time to copy them. They usually draw connections for you and lead you to the appropriate answer.	Various formats are used, including lectures, tutorials, online learning, computer-based, laboratory or field work, and self-study. Lecturers do not neces-sarily stick to the textbook, or even cover all the work. They are unlikely to stop and wait for you to copy down information and expect you to write your own notes as they lecture; they also require you to come up with your own connections and answers.
Learning styles	Passive – you learn by under-standing information prepared by your teacher and then repeating it. The focus is on studying towards passing final examinations.	Active – you now need to think critically, learn independently, and develop argu-ments of your own. The focus here is on understanding and creating connections between ideas. You will still need to study, but it's for a different purpose.

Table 1.1. Continued

Characteristic	Secondary environment	Tertiary environment
Class size	*Usually smaller classes of under 40 students in which teachers know your name.*	Usually much larger classes that can be in excess of 500 students. Lecturers are unlikely to know who you are.
Feedback	*Frequent feedback from teachers is given to you and your parents in the form of oral feedback, parents' evenings, and report cards.*	Less frequent feedback is given, and you need to manage your own performance. An academic transcript may be the only formal measure of feedback.
Goal	*To prepare you for a final senior certificate examination that determines your entry into tertiary study.*	To develop your ability to build knowledge and skills, think critically and prepare you for the working world.
Freedom	*As a pupil, you are usually given little freedom by parents or teachers, and are obliged to submit assignments, study for tests and exams. Your academic and social time is usually carefully monitored. However, there is usually less need to look after yourself in terms of washing, ironing, and cooking!*	As a student, you are given much more freedom and are expected to exercise this responsibly. Studying, attending class, submitting assignments, and socialising are (usually) personal decisions. But there is usually also the need to cook, clean, and do the washing yourself, so these freedoms come at a price!

Adapted from Southern Methodist University, n.d. and University of New South Wales, 2018

This means that to be 'tertiary-ready', you need to become an *independent learner* – someone who is self-motivated, able to identify problems and find relevant information, and who is then able to apply knowledge and skills to solve these problems consistently. It also means that you need to be able to reflect *critically* on your personal learning and performance. The ability to do this will also prove invaluable when you reach the working world, where employers expect a great deal of cognitive flexibility in their employees.

As you will have seen from this section, the move to tertiary studies is going to require some dramatic changes, and will provide you with new but sometimes intimidating experiences. Complete the short 'My Study Environment' assessment on the Learning Zone then go on and watch two videos about what managers want from their employees (LZ 1.1).

1.1.2 Values, attitudes and skills for tertiary studies

International studies on education and employment (such as QS Intelligence Unit, 2017; Mourshed et al., 2013) claim that there is a disconnect between employers, educational institutions and young people. For example, the Mourshed study (2013) found that 72% of education institutions sampled believe that their graduates are work-ready after graduation, but fewer than 50% of employers and graduates actually agree that they are. Additionally, the more recent QS Intelligence Unit

Harambee Youth Employment Accelerator's CEO Maryana Iskanker shares some valuable advice for South African graduates in a tough economic climate, and emphasises how important both communication skills and resilience are. You can read the full blog post at: <https://harambee.co.za/its-all-about-employability/>

study (2017) reveals that graduates think creativity and leadership skills are most important in a job, while their employers actually identify entirely different attributes as most critical. This means that your huge jump between school and tertiary studies may well be followed by yet *another* big jump when you graduate and move to the workplace.

Before this depresses you, you should know that most of the workplace skills and attitudes that employers are looking for are the very same ones that are important for academic success. Researchers (QS Intelligence Unit, 2017; Mourshed, 2013) identify, for example, the following:

- a clear capacity for critical thinking and problem-solving;
- a positive work ethic and willingness to adapt;
- a willingness to work constructively in a team where necessary;
- strong written and oral communication skills;
- a resilient approach to challenges and criticism;
- a degree of computer literacy;
- a basic ability to analyse information and data
- a grasp of simple mathematical functions.

To do something **consciously** means to do so with deliberate purpose and awareness. (Make sure you do not confuse this with *conscience*, which refers to the 'inner voice' that tells you right from wrong).

The debate on whose responsibility it is to ensure that graduates have these skills will continue. But as a student in control of your own learning and future, it makes sense to work consciously towards developing these skills and attitudes – even if your educational institution does not require you to do so.

A **manifesto** is a published verbal statement of what you intend to achieve, how you view a particular issue, and how you plan on achieving your goals. A manifesto is usually created as part of a political party statement, but a study manifesto is useful because you can look back to it in order to keep you focused on your goals. You can also update and adapt it as your goals change.

These days, most educational institutions have academic and digital literacy programmes that can help you to foster these important academic and workplace skills. Try to engage with these programmes fully, apply the skills to your formal subjects, and incorporate them into your everyday life too. This will help to improve your understanding, and so may also improve your results, thus delivering both short- and long-term benefits. If you can work on developing these skills, your next big step – into the workplace – will feel a lot less like a dramatic leap from a cliff.

 For this section, it is essential that you try to build and practise the skills and attitudes that employers are looking for, as well as being able to recognise them. Before you move on to the next section, visit the Learning Zone, and follow the instructions for developing your own Study Manifesto (LZ 1.2).

As you can see from this section, there is a lot more to the move to tertiary than a heavier workload and more freedom. Understanding that different attitudes and values are needed will also help you to adapt to the different kinds of learning – some of which are dealt with in the next section.

1.1.3 Learning styles

Simply put, a *learning style theory* is an academic theory that tries to describe how individuals differ in the ways that they prefer to receive and process information in different learning environments. For example, a learning style theory might argue that Person X prefers to learn by *doing* something practically, while Person Y prefers to learn by *seeing* how something is done.

Made popular in the 1970s, many of these theories suggest that lecturers and teachers should try to use the learning styles of their students to teach, and only then assess learning. And while there is serious criticism of learning style theories (Coffield et al., 2004; Stahl, 2002), you may have noticed that you do prefer to learn in some circumstances more than others. For example, you may understand a concept better if it is presented to you visually (in a picture or graph) than if someone just explains it; or you may prefer to be left alone to read through work instead of trying to follow a lecturer droning on in the classroom.

However, critics from the neuroscience and education fields (Coffield et al., 2004) argue that even if we *do* have an obvious learning preference, it probably stems from previous experiences, and not from any cognitive differences. For example, if you had a particularly boring teacher in primary school, you may have 'switched off' your preference for listening – or, if you had a particularly engaging teacher, you may now prefer to listen to lectures, podcasts, or discussions to learn and understand ideas. In this case, you would not *be* an 'auditory learner', but you may well have an auditory learning preference.

Critics therefore believe (Coffield et al., 2004) it is more important to try to maximise your chances of receiving, understanding and processing information. The best way to do this may be by *varying your learning environments* and the *ways in which you try to learn*. And although it is important to identify which learning styles you prefer, you should be careful not to label yourself as one type or another – this may be more limiting or even damaging than not recognising your learning style preference at all.

Some of the more developed learning style theories (such as VARK Learn Limited, 2019) acknowledge the theoretical limitations of these theories. But understanding your learning preference is still useful: even some sceptics of learning styles (Petrie, 2014) argue that the act of completing a learning style questionnaire can help you to become more aware of your cognitive processes, and so is a valuable exercise in and of itself.

You may therefore benefit from visiting one of the more popular sites to complete a questionnaire yourself. A good example of a learning-style questionnaire that is free for educational use is the VARK (Visual, Aural, Read/Write and Kinaesthetic) site, which was developed by Dr Neil Fleming. This more holistic theory says that learners usually show preferences for more than one type of learning, depending on the situation, and so refers to multimodal learning.

Cognition refers to the act of thinking. It includes elements of awareness, perception, reasoning and judgement. For example, 'Tokozile had to use a number of advanced cognitive strategies to respond to the really tough exam question.'

A **sceptic** is someone who likes to ask questions about opinions, claims and knowledge that others may accept as truthful or factual. Scepticism is an important part of academia, and constructive scepticism should be seen as a positive. For example, 'Abed revealed his scepticism by asking the lecturer questions about the credibility of the study she cited.'

Multimodal learning simply refers to the idea that the more ways that you try to learn something, the more likely you are to really learn it. For example, 'Alex is a multimodal learner, because he learns by reading the text fully, engaging in class discussions, and then making mind maps of his work.'

 The VARK site's questionnaire provides useful, practical tips that may help you to improve your learning experience across all types of learning preferences. See https://vark-learn.com/.

Understanding the different approaches that you can take to learning new ideas, together with your learning style preferences, is an important part of building your **academic identity**. The next section will introduce you to some of the basic cognitive strategies that you will also use to build your academic identity. You will engage with these strategies in more detail in later chapters of this book.

Figure 1.1, which follows, provides you with a brief outline of the type of cognitive strategies that you will use in your tertiary studies.

Comprehension strategies will help you to understand different texts and problems, and also assist you with answering questions and problems correctly.

Corroborate means that you need to confirm something by double-checking evidence, or finding another source to support it.

Figure 1.1. General cognitive strategies

Comprehension strategies

》 Monitoring and reviewing material through reading and checking for understanding
》 Using text structure to analyse cues given by writers to help organise ideas as you are learning
》 Summarising helps to identify the most important concepts and express them in your own words
》 Elaborating requires you to find new information and link it to given information
》 Explaining means that you need to ask yourself 'why' questions, and so helps you to understand content

Problem-solving strategies

Polya was a Mathematics professor from Hungary who developed problem-solving strategies.
》 Polya's strategy uses four steps:
  1) Understanding the problem;
  2) Developing a plan to solve the problem;
  3) Implementing the plan;
  4) Evaluating the plan's success and taking any necessary actions.
》 Remember that problems can be complex, so setting sub-goals may be useful to break more complex problems into manageable chunks.
》 Examining contrasting problems and extremes is also an excellent way of solving problems — looking at a system that works, for example, can help you to identify what does not work!

Writing strategies

》 Plan — come up with a bank of ideas and ways to organise them logically.
》 Generate sentences using accepted grammatical structure and logical flow.
》 Revise — excellent writers revise throughout the process, and do so critically, often changing major parts of their work, if necessary.
》 Use knowledge transformation, which takes existing ideas and transforms them into new ideas and structures — avoid telling!

Reasoning strategies

》 Generate counterarguments that oppose your own — this allows you to consider different perspectives and helps you to tighten your own arguments.
》 Be fair-minded — be as critical of evidence that supports your argument as evidence that does not.
》 Consider relevant comparison or control groups.
》 Consider the sources of information when evaluating it.
》 Corroborate information by consulting multiple sources to check facts and arguments.

Adapted from Chinn and Chinn, 2009

Problem-solving strategies will be very useful for when you are given more complex tasks and are asked to find solutions or adapt different scenarios, for example.

Writing strategies are also vital to almost every part of your academic and postgraduate careers, and so it is essential for you to focus your efforts on these too.

Although you should use arguing, evaluating and critical thinking from the start of your studies, you will notice that the further you advance in your course, the more you need to use these skills. This is where reasoning strategies fit in. Although they may be more challenging than writing strategies, you will realise how important they are when you practise them in later chapters. The ability to understand, solve problems, reason and then report on what you have done all feature as critical cross-field outcomes in academic studies across South Africa.

Having reliable active listening skills will help you develop comprehension, problem-solving, and reasoning skills in your tertiary career too – they are essential for getting the most out of lectures. Most of us could use a little help with listening, so if necessary, head to Section 13.5 and start incorporating these tips (as quickly and consistently as possible).

Critical cross-field outcomes are the specific outcomes that the South African Qualifications Authority considers essential for students to develop a capacity for lifelong learning. These outcomes are worked into all South African qualifications, depending on the level of study and to varying degrees. For example, 'Najwa Michaels, who lectures Mechanics 101 on an Engineering degree, focuses on teaching her students how to use technology responsibly in the environment throughout her course – of all the critical cross-field outcomes, she thinks that this is the most important focus of learning.'

1.1.4 Metacognition

In addition to cognitive strategies, one of the most important elements of cognition is the act of thinking about **why** or **how** you acted or thought. This is called metacognition, and can cover a range of events, from a simple awareness that you understand the content of a subject more quickly or more slowly than others do, to more complex aspects, like understanding why a particular argument you have made is (or is not) academically sound. Although this reflection, or metacognition, is often neglected, it is one of the most valuable kinds of cognitive skills that is available to us and you will practise it frequently throughout this textbook. As Diderot points out (2000), 'There are three principle means of acquiring knowledge ... observation of nature, reflection, and experimentation. Observation collects facts; reflection combines them; [and] experimentation verifies the result of that combination.' If you want positive outcomes in your 'experiments' in life and studies, then do not forget to combine study with reflection.

Biographical note

Denis Diderot
(1713–1784) was a French Enlightenment-period philosopher, writer and art critic.

The activity below incorporates much of the content and practice of this chapter as a whole. It should assist you to consolidate your understanding of what tertiary studies entail by asking you to engage your metacognitive skills. Think about the scenario below (adapted from Tanner, 2012) and then respond to the questions that follow:

Xolani is a first-year BCom student and has arrived in his Economics lecturer's office the day after his first test in the subject. When his lecturer, Lerato, asks him how he prepared for the test and how it went, Xolani tells her that he was glad that it was on a Monday, as it gave him plenty of opportunity to study – straight through from Friday afternoon to Monday morning. He tells her proudly that he did not go out with his friends at all over the weekend, and that he read, then reread all the test-work and even made flash cards of the words in bold in his textbook. Because of all his effort, he feels that he should have done really well in the test, but for some reason, he feels very anxious about his results.

Sarah, another of Lerato's Economics students, also visits her office the day after the test. When Lerato asks her how she prepared for the test and how she thinks it went, Sarah says that she is not too worried about her test result as she spends time every afternoon going over her class notes and PowerPoint slides, and then comparing these with the textbook content. She tells Lerato that there were quite a few points that seemed to contradict each other which confused her, but that she discussed these points with her classmates on her class's Online Discussion Forum. She says that these discussions resolved most of her confusion and made her think about the work more. After writing the test, she realised that there seemed to be a few areas that she might have misunderstood or missed out, so she has some questions she would like Lerato to help her with so that she can correct her understanding for future assessments.

Based on your understanding of the demands of tertiary studies, write a paragraph explaining which of these two learners you think will:

a) Have learnt (according to tertiary study objectives) the most from studying for the test?

b) Obtain the better mark, assuming that both students are equally bright?

If you are able to, discuss your answers in small groups (either online in your class forum or physically, in class). Once you have considered your responses, use short paragraphs to reflect on *why* you answered the way you did by asking yourself the following questions:

• How did you arrive at your decisions? What elements did you consider?
• Did your classmates arrive at the same conclusions?
• What arguments did you not think about, or did you consider but decide to dismiss?

1.2 Conclusion

This chapter introduced you to some of the key differences between school and tertiary-level studies, and helped to familiarise you with the values, attitudes and skills that you will need to succeed over the coming years. It also offered a short summary of different learning styles and cognitive strategies that will be developed in more detail throughout this textbook. Once again, welcome to your studies! We hope that you now feel ready to engage deeply!

References

Chinn, C. and Chinn, L. 2009. 'Cognitive strategies'. *Education*. Accessed 23/08/2019, available at <https://web.archive.org/web/20161223162530/http://www.education.com/reference/article/cognitive-strategies/>

Coffield, F., Moseley, D., Hall, E. and Ecclestone, K. 2004. *Learning styles and pedagogy in post-16 learning. A systematic and critical review*. London: Learning and Skills Research Centre.

Diderot, D. 2000. *Thoughts on the interpretation of nature and other philosophical works*, translated by Sandler, L. Manchester: Clinamen Press.

Iskanker, M. 2019. 'It's all about employability'. *Harambee Youth Employment Accelerator*, 3 June 2019. Accessed 23/08/2019, available at <https://harambee.co.za/its-all-about-employability/>

Mourshed, M., Farrell, D. and Barton, D. 2013. *Education to employment: Designing a system that works*. McKinsey Center for Government.

Petrie, D. 2014. 'The learning style debate'. *TEFLGeek*. Accessed 23/08/2019, available at <http://teflgeek.net/2014/02/10/the-learning-style-debate/>

QS Intelligence Unit. 2017. 'The global skills gap: Student misperceptions and institutional solutions'. *Reimagine Education*. Accessed 20/08/2019, available at <https://www.reimagine-education.com/wp-content/uploads/2018/01/RE_White-Paper_Global-Skills-Gap-Employability.pdf>

Southern Methodist University. n.d. 'How is college different from high school'. *Student Academic Success Programs*. Accessed 23/08/2019, available at <http://www.smu.edu/Provost/ALEC/NeatStuffforNewStudents/HowIsCollegeDifferentfromHighSchool>

Stahl, S.A. 2002. 'Different strokes for different folks?' In L. Abbeduto (Ed). *Taking sides: Clashing views on controversial issues in educational psychology*: 98-107. Guilford, CT: McGraw-Hill.

Tanner, K. D. 2012. 'Promoting student metacognition'. *CBE Life Sciences Education*, 11(2): 113–120. Accessed 23/08/2019, available at: <http://www.ncbi.nlm.nih.gov/pmc/articles/PMC3366894/>

Tyson, N. 2012. 'I am Neil deGrasse Tyson'. *Reddit Ask me anything thread*. Accessed 18/02/2020, available at <https://www.reddit.com/r/IAmA/comments/qccer/i_am_neil_degrasse_tyson_ask_me_anything/c3wgffy/>

VARK Learn Limited. 2019. Accessed 20/08/2019, available at <http://vark-learn.com/using-vark/using-vark-in-research/>

Vincent, L. and Idahosa, G. E. 2014. '"Joining the academic life": South African students who succeed at university despite not meeting standard entry requirements'. *South African Journal of Higher Education*, 28(4): 1433–1447.

University of New South Wales. 2018. 'Getting started: Tips for high school leavers: Differences between uni and high school'. *UNSW Getting Started*. Accessed 20/08/2019, available at <http://www.gettingstarted.unsw.edu.au/tips-high-school-leavers>

Was, C. A., Al-Harthy, I., Stack-Oden, M. and Isaacson, R. M. 2009. 'Academic identity status and the relationship to achievement goal orientation'. *Electronic Journal of Research in Educational Psychology*, 7(2): 627–652.

CHAPTER 2

Reading and referencing skills

'A man's mind is stretched by a new idea or sensation, and never shrinks back to its former dimensions'
– Oliver Wendell Holmes Sr (2013: Chapter XI)

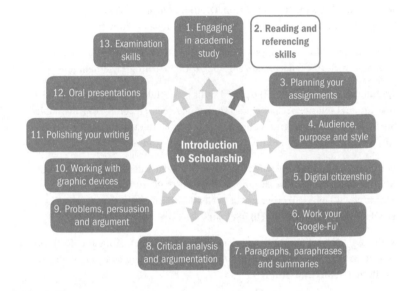

13. Examination skills

1. Engaging in academic study

2. Reading and referencing skills

12. Oral presentations

3. Planning your assignments

11. Polishing your writing

4. Audience, purpose and style

Introduction to Scholarship

10. Working with graphic devices

5. Digital citizenship

9. Problems, persuasion and argument

6. Work your 'Google-Fu'

8. Critical analysis and argumentation

7. Paragraphs, paraphrases and summaries

OBJECTIVES

At the end of this chapter, you should:

- understand the main reasons for reading;
- demonstrate familiarity with the purpose and structure of different academic texts;
- recognise the need for different reading strategies;
- employ suitable reading strategies for various academic purposes;
- understand the importance of academic referencing;
- correctly and consistently use the prescribed in-text and bibliographic referencing techniques.

IRL

Three weeks into the first semester of his BA degree, Andile is already feeling the pressure – he has no idea how he will get through all the prescribed material, complete all his assignments and tests, *and* then pass his exams. His first-year English modules alone have created a pile of 11 books on his desk! He has no idea how he will manage the work for all his other modules too.

His best friend Kamini, in her second year of studies, keeps telling him not to freak out and that he does not need to study everything in detail. Andile is unconvinced, worrying that he is going to disappoint his entire family by failing his first year spectacularly.

2.1 Introduction

Like Andile, you may be anxious about the volume of work you have to get through in a single semester. If so – welcome to the life of a student. There is no honest way to airbrush your concerns – they are legitimate. In fact, if you are not yet concerned about the volume of work, you may simply not have had a good enough look at what you need to accomplish.

However, you should not be so concerned by the size of the **curriculum** that you feel overwhelmed. Yes, tertiary studies are challenging. They require hard work and a huge amount of reading. They are designed to push your brain further, harder and faster. In fact, the very word 'curriculum' stems from the Latin word *currere*, which means 'to run' (Oxford University Press, 2020). *But* curricula are also designed around a formula related to **notional hours** so that your academic workload is kept within reason.

Tertiary institutions must use these notional hours when they are designing their courses and workloads. For example, your first-year modules should add up to approximately 120 credits, so you are likely to spend about 1 200 hours studying, attending lectures, reading, and doing assignments for all your classes in your first year. With a three-year degree set at 360 credits, you will need to spend about 3 600 hours studying before you graduate!

Credit-based notional hours is not an exact science though, so it may take you more work to pass all your subjects, especially if you are working towards passing well. The type of subject, the ease with which you understand the work, the suitability of your selected reading strategies, and your skills and reading speed will also affect the amount of time you spend on a module, so do not expect to use notional hours as a hard and fast rule. However, being aware of notional hours and how they are used should help you to plan your timetable more realistically.

This chapter will introduce you to different types of texts, and to different types of reading strategies and techniques that you may use throughout your studies. These are important for a number of reasons.

Firstly, you will save time by being able to recognise the type of reading to apply to different tasks, such as studying for a test vs reading to find information for an assignment will help you to save time. For example,

Curriculum, as used here, refers to all the knowledge and skills that you are expected to learn in a particular course or degree in order to meet the course objectives. These may include, for example, all the learning units, lectures, assessments, and prescribed material of a particular subject, or of the whole degree (Great Schools Partnership, 2015).

Notional hours refers to the total number of hours that the average student would need to spend on all learning activities (reading, attending class, completing assignments and studying), in order to achieve a pass mark in a course. The notional hour-spend is worked out at 10 notional hours per credit. For example, 'My first semester Psychology 1 module is worth 12 credits, so I should be able to get through all my work and pass if I spend 120 hours on it this semester.'

Note!

1200 hours may sound like a lot of time, but it actually amounts to just 50 (solid, 24-hour) days of work, so technically you would still have 315 days of the year to eat, sleep, work and party! Realistically, it's impossible to 'squash' your studies up like this, so rather work steadily and keep a life-balance to maximise your chances of doing well.

you do not want to waste time study-reading an entire chapter if you simply need to identify and study its main points.

Secondly, matching the right kind of reading strategy to a particular task will help you to engage in an appropriate cognitive process. For example, if you need to find the outcome of an election in a news article, you only need to skim the article for the result – but if you need to evaluate the findings of a journal article, you will need to use a more critical approach to your source, engaging all of your higher-order cognitive faculties.

Thirdly, from a philosophical perspective, consider the point made by educator and philosopher Paulo Freire about the enormous power of reading. In his book *Pedagogy of the oppressed* (1970), Freire argues that a critical literacy helps us to actively interpret and critique the world so that we can change it for the better, as opposed to the more common practice of merely describing it and accepting things as they are (Roberts, 1996). Simply put, Freire claims that being able to read critically frees you – firstly, by 'conscientising' you, or making you aware; secondly, by helping you to take informed action on what you have learned (Freire, 1970). Being a conscious, critical and reflective reader will therefore help you to engage fully in the business of being a 'human-doing', and prevent you from just passively 'being'.

Okay, before you feel we are getting lost in fluffy 'freedom' philosophy here, there is another important, and completely practical reason to want a toolbox of effective reading skills: being able to read the right text in the right way will also help you to pass your modules, and therefore help you to avoid prickly conversations about poor results with those wonderful people who pay your tuition fees! In fact, effective reading (together with being able to write clearly) will be one of the key determinants of your success academically.

....

'Reading is not walking on the words; it's grasping the soul of them' (Freire, 1985:19)

....

Biographical note

Freire (1921–1997) was a Brazilian teacher who worked to improve literacy in disadvantaged communities through his use of a **critical pedagogy**. Critical philosophy tries to raise awareness of unjust power-relations in society through education, and in Freire's case in particular, through the teaching of critical reading skills (Roberts, 1996).

To **cite**, in this context, means to ensure that your sources are recognised throughout your work, both through in-text and bibliographic referencing.

Right, so now you understand *why* you need reading skills. But how are you going to cite what you have read in a way that is consistent with your faculty's rules and regulations?

The second part of this chapter will focus on how to *correctly* and *consistently* use in-text and bibliographic referencing in the style that your faculty prescribes. It probably does not sound like much fun – but you need to use referencing in *all of your assignments* – right from the very start of your studies, so it is worth learning. If you don't reference properly, you risk being awarded zero for your hours of blood, sweat and tears; you could even receive a warning from your university on plagiarism infringement. This is not the way you want to start your university career. So, focus on this chapter, and promise yourself to practise the reading and referencing techniques as often as you can. From our side, we will try to make mastering citations as painless as possible!

For this chapter, you will find on the Learning Zone separate reference guide examples and links for some of the main styles of referencing used in southern African universities. These include the Harvard method (used in most humanities and business faculties), the American Psychological Association (APA) style (used by psychology students), and the Vancouver style (used by many science and health science faculties). There isn't currently a single official citation style for law faculties in South Africa, but a guide for law students based on the *South African journal of human rights*' in-house style is also provided in the Learning Zone.

2.2 Academic texts: Purpose and structure

There are a number of elements to consider when approaching an academic text; two that will provide valuable context are the *nature of the text*, and your *reason for reading* it in the first place. This section will look at some of the different types of texts you might encounter, as well as some of the main purposes for reading them.

2.2.1 Types of academic texts

Although all students will need to master strategies for textbook reading, different qualifications will expose students to different kinds of academic texts. For example, a Bachelor of Arts student taking English will have to read the textbooks, news articles, graphic devices and journal articles that other students encounter in their studies, as well as critically reading their set texts, such as novels.

Note!

Annotated samples of each kind of text mentioned in Table 2.1 are available on the Learning Zone (see LZ Student exercise 2.1), together with links to relevant videos. Please consult these samples to get a better idea of the appropriate reading approaches for, and overall structure of, each type of text.

Table 2.1 outlines the main structural features and functions of some of the texts you are most likely to read for your studies. Your particular text may not have all of the structural elements listed, or it may feature a few different ones, but the table shows the most common features.

Table 2.1. Structural features and purposes of select academic texts

Type of text	Typical structural features	What this feature does
Textbook	Table of contents	Gives you a simple overview of what the book covers in just a few pages – some textbooks provide two kinds: a 'simple' and a 'detailed' table of contents
	Preface	Establishes the importance and/or credibility of a book through an introduction, often by an external reviewer
	How to use this book	Describes how to approach the reading of the book as a whole, and what to do with particular or unique features
	Chapter title	Tells you what a chapter is about

Table 2.1. Continued

Type of text	Typical structural features	What this feature does
	Chapter objectives	What you need to be able to do once you have read a chapter
	Introduction	What is to be covered in a particular chapter
	Subheading	Explains what question will be answered – usually relate to a chapter objective
	Visual	Could be a map, chart, diagram, graph, table, cartoon, photograph etc. If well-designed and well-used, a visual can help create a more complete understanding of a topic through giving examples, showing processes, making comparisons, enumerating lists, and classifying or generalising concepts (Leavitt, n.d.). Should be accompanied by a caption and/or title that shows how they fit in to the content
	Key term	Describes or defines new vocabulary or jargon
	Specific typography	The use of different fonts, highlights, colours, weights (e.g. bold), text sizes etc. emphasises or categorises text to draw attention to certain words or phrases
	Icon set	A set of specially developed images that alerts the reader to different types of content, such as activities, word-use notes, skills-building exercises, reflection ideas etc. For example, in this book, icons are used to alert you to digital resources, additional reading ideas, etc.
	Margin note	These are notes or textboxes in the margin of the page, including references to activities, information about key terms, or references to interesting links
	Summary and/or conclusion	Provided at the end of a chapter to consolidate what was covered in the chapter
	End of chapter set of questions	Test your understanding of what was covered in the chapter – sometimes these are online instead of in the printed copy
	Glossary	An alphabetised list of definitions or descriptions of terminology and jargon introduced in the book. Sometimes combined with the index to create an indexed glossary
	Index	Appear in most good textbooks – an alphabetised list of important terms or concepts with page references where the terms or concepts are used in the book. Very useful for locating information quickly
Graphic device	*Table*	Used to organise and present information, perform calculations, and to highlight important parts of specific information for the reader (English, 2012)
	Line graph	Used to show continuous information or trend curves to professional audiences
	Bar graph	Shows comparisons of two or more items, or component parts of these items (English, 2012).
	Pie chart	Shows relative percentages that make up a whole
	Pictogram	Uses icons to show statistical information to a broader, and possibly not professional, audience (English, 2012)

Table 2.1. Continued

Type of text	Typical structural features	What this feature does
	Flow chart	Useful for problem-solving, decision-making, and for showing the steps in a procedure
	Gantt chart	Used for project management – provide a visual representation of the sequence and current status of tasks in a project
Journal article	Title	Briefly describes the focus of the paper, giving the reader a clear picture of what to expect from it
	Author information	Includes name, institutional affiliation etc.
	Journal information	The title of the journal, its volume, issue and relevant page numbers
	Abstract	Summarises the purpose of the study, the research design method, the major findings and the conclusion
	Introduction	Provides a context, general overview, and describes the specific questions that the article will examine
	Literature review	Gives an overview of what other authors have said on the topic
	Methodology	Describes how the research was carried out or how information was collected and analysed
	Results	Objectively reports on the key findings, without interpreting them, in a logical sequence, using text, tables, and figures
	Discussion	Interprets the results in light of what is already known about the topic, and answers the questions posed in the introduction
	Conclusion	Summarises the findings
	Reference section	Includes citations of all the authors referenced in the article; may be in footnote or bibliographic format
News article	Headline	A short, usually attention-grabbing statement that summarises the story
	Byline	The name of the person who the story
	Lead paragraph	Usually outlines the 'who, what, when, where, why, and how' of the story (although there are different styles)
	Explanation	Includes the details, quotes and other important information about the story or event
	Additional information	Adds information about similar or related events, but is not essential to the story
Literary work	Novel	Primarily used to entertain; describes the human condition, using prose, chapters, and a typical 'setup-conflict-resolution' structure
	Poetry	Poetry uses stanzas, symbolism, metre and rhythm to convey emotion, describe imagery, evoke responses, comment on society, protest against it, or tell a story
	Drama	Otherwise known as a play; uses scripted dialogue between characters – performed in theatres or similar

2.2.2 Reading purposes

"It's called 'reading'. It's how people
install new software into their brains"

Before starting to engage with a text, it is important to consider why you are reading it. For example, an actuarial science student may pick up a novel to relax but a student majoring in English will need to approach a set text such as a novel more methodically. An Economics101 student trying to find an interest rate percentage in a news report will also use a different reading strategy to someone examining the reasons for an economic recession in a journal article.

Twelve important reasons for reading

There are other reasons for reading, but most people read because they:

  1. wish to 'switch off' and relax;

A **repertoire** is a set of skills or behaviours that someone can, or does, do. For example, 'Xolile's cooking repertoire is really limited, consisting exclusively of macaroni cheese and two-minute noodles.'

The word **ibid**. is a contraction of the Latin word 'ibidem' meaning 'in the same place', and is used to cite sources that have been cited immediately before. In this case, for example, the use of *Ibid*. means that the previous cited source in this paragraph, Winter (2019), is also the source of this information.

Point of interest

Reading to 'switch off' and relax can contribute significantly to your learning (ahem, Twitter-feeds and Facebook updates don't count). Even though leisure reading will not directly contribute to your academic success, keeping it in your *repertoire will* extend your vocabulary, stimulate your brain functioning, develop your writing skills, build empathy, improve your focus and concentration, cultivate analytical thinking skills, and broaden your worldview (Winter, 2019). This applies to first-language English speakers, but students with English as a second language (ESL) will benefit even further from reading leisure books in English, since seeing words in context can help to improve both speaking and writing fluency (**Ibid**.).

---

  2. are interested in a particular subject;
  3. want to find out what an author thinks on a topic;
  4. need to identify the main elements of something;
  5. need background information;
  6. need specific information that is likely to be in the text;

7. are writing an essay on the topic;
8. need to present a comprehensive argument for or against something;
9. need to persuade someone of something related to a topic;
10. have to argue for or against what an author has claimed;
11. are going to have a discussion on a subject;
12. will be formally examined on the content.

Remember these reasons, and ensure that you first ask yourself *why* you need to read a text, before you start reading it. If you are unsure what your purpose is in reading a prescribed text, go back to the instructions you were given on the text, or look at the subject outcomes or objectives. These should tell you why you need to read the text, and also the depth of reading required.

For example, let's assume that you are preparing for an Economics class by going through your subject's study guide. You are instructed to read Chapter 4 of the guide to ensure that you are: a) able to identify macroeconomic factors that affect business performance; and b) that you are able to debate the benefits and drawbacks of minimum wages. In the first case, your purpose is to identify the key elements of something, while in the second case your purpose is to present a comprehensive argument for or against the topic. Your reading for the first task would therefore certainly not be as extensive as it would be for the second task, leading you to select different reading strategies for the tasks.

For most instructions, you should be able to find your answer in the list of 12 important reasons for reading above, but don't panic if you can't – there are hundreds of good reasons for reading. Just aim to figure out what yours is for the text that you are working with so that you select a suitable approach.

2.3 Purpose-based reading strategies

Okay, so now you know *why* you need to read your chosen text. The next step is to select a reading strategy that will help you to achieve your purpose. This section will outline five main reading approaches as well as outlining specific techniques to use to achieve your purpose. Specifically, we will look at light reading, scanning, skimming, reading for overview, intensive reading, and study reading. Table 2.2 below classifies these reading approaches and describes the purposes that each approach is suitable for. Some strategies are suitable for more than one purpose, so think carefully about which option will work best for each text that you need to read.

 Use student exercise LZ 2.2 on the Learning Zone to round off your understanding of reading strategies and do a quick check of how to apply them.

Table 2.2. Reading approaches for specific purposes

Reading approach	Best if you:
Light reading	• want to 'switch off' and relax; • are interested in a particular subject.
Scanning	• need to find specific information in a text.
Skimming	• want to find out what an author thinks on a topic; • need to identify the main elements of something; • need background information.
Intensive reading	• need to analyse a literary extract in fine detail to understand it fully.
Study/Critical reading	• want to find out what an author thinks on a topic; • are interested in a particular subject; • are writing an essay on the topic; • need to present a comprehensive argument for, or against something; • need to persuade someone of something related to the topic; • have to argue for or against what an author has claimed; • will be formally examined on the content.

Now that you have an idea of a suitable reading approach to use for a particular purpose, we will examine all five reading approaches in the next section.

2.3.1 Light reading

As shown in Table 2.2, we tend to use light reading when we are reading for leisure. When we use light reading, we usually:

- choose content we find interesting;
- read at an unhurried pace of between 100 to 200 words a minute (Kruger, 2004);
- focus on grasping meaning;
- become emotionally involved in the text;
- skim over boring, irrelevant sentences, paragraphs or passages;
- don't worry too much about concentrating.

For this kind of reading, the only strategy you need is this: find a book that looks interesting, put your phone on silent, get comfortable, and open the book. The writing should invite you in, and you should enjoy a relaxing, world-expanding form of reading.

Before literature students get excited about how 'easy' their studies are going to be, remember that you will not be using this reading approach for your set texts even though you will be reading novels as part of your studies. Reserve this approach for weekend reading, please.

Have a look at Modern Library's 100 Best Novels – www.modernlibrary .com/top-100/100-best-novels – or head over to Goodreads – www.goodreads .com – which uses social media and the 'Listopia' concept (a list of recommended books that meet defined criteria) to guide readers to books that they will find interesting .

2.3.2 Scanning

As a reading approach, scanning is a limited activity in that you look exclusively for information that is relevant to your search, and ignore everything else. However, it is very useful for finding information quickly, and in academia you might use it to kick-off a research project. For example, if you did a Google search on an assignment topic, you would scan the list of returned items to see if any were relevant, without spending time reading every result. Automatically you would be scanning the results: you have probably been scanning for years without being aware of it.

Here is a simple technique to hone your scanning:

1. Figure out what exactly it is that you are looking for (i.e. establish your *purpose*) and have it clear in your mind. Either focus on a keyword, or on a symbol, depending on what you are looking for. For example, if you are looking for a percentage increase in a news article, keeping the '%' sign in your head would help you find the increase.

2. Then, work out the *structure* of the information you are reading. For example, is it alphabetised, arranged **chronologically**, by category, by order of importance, or just in paragraphs? Think about the usual structure of the type of text you are working with, whether essays, news articles, textbooks or other items, and identify where the information you need is most likely to be.

3. Now keep the *image* of the word or information you are looking for in your mind, and use your *finger* to guide your eye down the page you think is most likely to feature the information you want.

> If something is arranged **chronologically**, it means that the items are displayed in order of their occurrence in time. For example, a historical timeline would be arranged chronologically from the oldest event to the most recent, whereas a CV is often organised from the most recent to the oldest event.

Remember that this is not a thorough reading technique, and you should only use it for specific purposes, such as figuring out whether a text includes the information you need.

2.3.3 Skimming

 When you have read Sections 2.3.2 on scanning and 2.3.3 on skimming, use student exercise LZ 2.3 on the Learning Zone to check and reinforce what you have learnt.

Skimming is a quick way to find the main idea or points of a text (Du Toit et al., 2000). It is useful if you want to review something you have already studied or read in-depth, if you want to look over a text before you read it more carefully, or if you need to answer a particular question and know what kind of information you need.

When you skim-read, you skip over a substantial portion of the text; to avoid just randomly throwing your eyes across a page and landing on **arbitrary** nonsense, we advise you to work methodically.

> **Arbitrary** refers to something being haphazard, random, or by chance. For example, 'The arbitrary inclusion of that sex scene in the movie made no sense at all.'

To skim-read in a way that is likely to improve your chances of finding the main points, try the following:

1. Before you begin, know what your purpose is and keep it in mind.
2. Quickly check the *title and introduction* of the text and ask yourself what you already know about the topic.
3. Now run your eyes over the text and look for *subheadings* or words in bold or italics that might provide you with important information.
4. Next, check the first sentence of each paragraph. This is usually the *topic sentence*, which describes the main point of the paragraph, allowing you to skip the remainder of each paragraph. Do this for the rest of the text until you get to the final paragraph.
5. The last paragraph is likely to be a *summary* of what you have read. Read this more carefully than you read the main text to check that you covered all of the main ideas in your skim of the rest of the text.
6. Once you have finished skimming the full text, *ask yourself* if you achieved your purpose – if not, repeat the process; if you did, well done – you have saved yourself some time.

The THIEVES technique is another short but effective approach you might want to try. THIEVES is particularly useful for skimming through a textbook. Hatzi (2009) suggests that you skim the:

- **T** = Title
- **H** = Headings
- **I** = Introduction and first paragraph
- **E** = Everything you know (how does this relate to what you know already?)
- **V** = Visuals (what do the images tell you?)
- **E** = End of chapter questions (do these suggest the main themes you skimmed?)
- **S** = So what? (how does this affect your understanding of the subject as a whole?)

As we pointed out earlier, please do not use skimming as a study technique. Although it is useful to review work that you have already studied, we would not bet on your success if you use this method to study!

 Practise your 'thieving' on the Learning Zone: use student exercise LZ 2.4.

2.3.4 Intensive reading

In comparison with the previous three reading approaches, intensive (or close) reading is incredibly thorough and cognitively demanding, focusing on comprehension and retention over a long period. Literature students, in particular, will need to be skilled in this kind of reading in order to analyse shorter pieces like poetry, or extracts from screenplays, speeches or novels.

In essence, intensive reading zooms in on the various elements of a piece of writing, such as literal and figurative meanings, the literary structure, the use of rhetoric, tone and emotive language, and even the presence of social commentary or inferences. This technique relies on your understanding of vocabulary, and literal and figurative meanings, so it takes time, determination, and concentration to master (coffee may help too).

You may have used aspects of this technique at high school when you engaged with the poetry or drama elements of your home language subject; perhaps you felt like you were 'picking' apart these works. Don't worry if your skills are rusty – the tertiary studies version probably requires a more critical eye than your initial practice anyway – just try to practise the technique as frequently as possible. This is especially important if you are studying humanities or social sciences, where you will need to critically analyse texts of various kinds.

Although some scholars (see MacLeod, n.d.) argue that this is a skill best practised in class and led by a lecturer, you can move at your own pace and improve your comprehension by practising the technique below in your own time. Ideally, you can buddy-up with a classmate or online forum member and share your understanding once you have completed an exercise.

Rhetoric refers to a group of literary devices, including figures of speech, that are often used in literature or speeches to influence or persuade readers or listeners. A good example is Martin Luther King's 'I have a dream' speech, which made use of a rhetorical device by repeating that one sentence ('I have a dream') at the start of a series of paragraphs.

Point of interest

You should recognise most of the literary terms below from high school, but if you do not, don't despair – you will be using them frequently throughout your studies so these will become regular features of your adventures in analytical reading.

Here is one version of an intensive reading technique:

1. Ensure that you have a clear *purpose* in mind – ask yourself:
   > What do I need to be able to do with this text?
   > Is there an element of the writing I need to focus on?
   > How detailed does my comprehension need to be?
   > Will I need to recall this information for future use?
   > What do I already know about the text and its author/s?
2. Use the *scanning* technique to read through the text once to find the main theme and identify the different structural elements of the text.
3. Write down any *questions* that occurred to you on your initial reading.
4. Reread the text *line-by-line* carefully and thoroughly, focusing on comprehending the ideas in each one.
5. Read the text again slowly and try to *identify* the following:
   > the genre of the writing;
   > how the text is structured (i.e. look at its form);
   > the narrative voice;
   > the main theme/s or concepts (make sure that you express these in your own words!);

The word **genre** is used to describe different types of artistic compositions that share similarities in form, subject matter or style. For example, 'Some of the classic genres of literature are comedy, tragedy, and romance. Lately, the fantasy and sci-fi genres have become very popular.'

The **narrative voice** refers to who 'tells' the story in a piece of writing, i.e. it is the 'voice' that speaks to the reader, and is important to consider in terms of reliability. Narrative voice can be first person (in which case the speaker would use 'I') or third person (in which case the speaker would use 'he'/'she' or other, more impersonal ways to reveal the story). For example, 'Most of the journal articles I have read make use of the third-person narrative voice to make their research look more objective.'

> the tone, or changes in tone;
> the literal argument;
> figures of speech such as metaphors, allusions;
> evidence of rhetoric;
> the figurative meaning or intended implication;
> how the text makes you feel (yes, this is important too).

6. Once you have explored the text as much as you think you can, *test* your understanding by either writing a short summary, or by discussing your understanding with a classmate or student on an online forum who has read the same text.

2.3.5 Study/Critical reading

Study/critical reading is the final reading approach we will look at in this chapter; it is also the one that you are likely to use most frequently, but perhaps enjoy least. Sadly, you *do* need to be able to study if you want to pass. The good news is that using a solid study reading method means that you can improve your understanding of the content, so you will not need the old-fashioned 'parrot-studying' you may have used at school.

Remember that study reading can be used for a number of purposes, including critiquing or examining texts, and not just for studying (see Table 2.2 for a reminder of these purposes). The SQ3R method of study reading is effective and popular although it is not the only method. SQ3R stands for *survey, question, read, recall,* and *review*.

Whichever study reading method you choose, the goal should be to develop yourself into a **lifelong learner** – you are not here just to learn 'stuff'. As a result, it is important that you focus on developing critical, or 'deep', reading skills. You can do this by:

- approaching texts with the intention to understand what they say, as well as how they say it, and what they imply;
- noting important elements of a text such as the author's purpose, use of tone and persuasion, and the presence of any **bias** in the text;
- relating the content of the text to what you already know or have experienced;
- evaluating the quality of evidence to decide whether or not an argument is logical and supported.

If you think about how you read academic texts and can already tick all four bullet-points above, then you definitely have a head start on comprehension and critical reading. Most of us, however, fall into the category of 'surface readers' and tend to see most of what we read as facts (or true) without asking many questions. We need to work to develop our critical reading skills. Regardless of whether you are a 'deep' or 'surface' reader, practising the SQ3R technique will improve the level at which you read, understand, and remember texts, and so it will help to improve your chances of success, regardless of your reading skill.

Lifelong learning describes an attitude to personal growth that embraces different kinds of learning at different times and places in life. Lifelong learners do not simply wish to master their fields of study to get a certificate: they also want to learn how to implement their understanding practically, and are inclined to keep learning beyond what they are required to do. This makes them very valuable, flexible employees in a world that is perpetually changing.

Bias is an inclination towards, or prejudice against, one argument, idea or person, which benefits another argument, ideas or person. For example, 'That referee was so biased – the other team never had a chance.'

The SQ3R method might look like a laborious way to go about understanding and remembering your work, but if you can turn it into a habit, you will probably find that you need to read most texts fewer times – making it worth spending some time mastering the technique.

As with all the methods we have covered, it is important to understand your *purpose* for reading before you begin to use this method. If you don't know what you are trying to achieve, you may waste time.

In step-by-step format, the SQ3R requires that you:

1. *Skim* (or *survey*) the text. In this step, you need to get an overview of the:
   > title, headings and subheadings;
   > images, graphs, charts and accompanying captions;
   > introduction;
   > first sentences of each paragraph;
   > conclusion or final paragraph;
   > review questions, or additional notes, if any.
2. *Question* the text:
   > convert the title, headings and subheadings into questions;
   > read the questions at the end of the chapters;
   > now look at all of the questions you have identified, and ask yourself:
     - What did my lecturer say about this content?
     - What do I already know about this?
     - Is there criticism against this content?
     - Is this information part of one of my course outcomes or objectives?
3. *Read* the content *actively*, one section at a time, by making notes, or annotations on the text itself (for study reading, also recite from memory the key ideas in your notes before moving to the next section):
   > as you read, try to find answers for the questions from Step 2;
   > slow down your reading speed when you get to more difficult sections – identify what you do not understand by putting a question mark (?) in the margin;
   > circle the author's main points, and note down 'MP' in the margin;
   > reread any images, graphs, charts and captions;
   > take note of words or terms in **bold** or *italics* and <u>underline</u> any unfamiliar terms;
   > highlight important evidence or arguments by using a highlighter or putting an asterisk (*) in the margin;
   > identify statements that you react strongly to by making margin notes or using an exclamation point (!) in the margin;
   > identify related points by numbering them in the margin.
4. *Recall* (and write) what you have covered after each section:
   > don't 'parrot' everything you have read – you need to summarise the content in your *own words*; if possible, do it out loud (yes, people will think you are nuts, but the benefit of putting into words what you have read and hearing it said is worth it);

> ask yourself oral questions about the content – see if you can answer those questions you identified in Step 2 – go back to the text if you missed something;

> write your notes once you feel you can comfortably summarise the section and answer questions *without looking at the text*.

5. *Review* the content and your notes:

> reread the section and your notes – check for inaccuracies;

> cover the text in your book and look only at your notes – do you know what each of your notes refers to?

> ask yourself these questions:
   - Can I identify the main point or problem?
   - What solutions are offered?
   - Has the author provided fact, well-supported theory, or opinion?
   - Is there evidence for claims made?
   - Has the author adequately proven his/her point?
   - How would this affect daily life or practice?

> If at all possible, explain and debate your take on the material with a classmate who has covered it – this will help to contextualise the ideas, making you more likely to understand and remember it in the long-term.

Note!

If you are using this technique for study purposes, Step 5 is an ongoing step and should happen immediately after you have made your notes, the day after, and then at decreasing intervals until your test or exam to allow your brain numerous opportunities to interact with the content. Regular reviews will also help to embed the material in your long-term memory increasing your ability to recall it.

 Student exercise LZ 2.5 on the Learning Zone will help you explore how this strategy works in practice, and how it works for you.

2.4 Why we need to reference

Most students struggle to adapt to referencing in their first year of study. This is because referencing is often a new concept to students, since not all schools insist on the practice. If it is a new concept, the problem is that referencing is usually left to the final chapters of first-year academic literacy textbook. So, by the time you get to practise referencing techniques, you are likely to have already handed in many assignments and learnt your lesson the hard way, by getting low marks.

Referencing skills themselves are also seldom covered in your formal subjects – instead lecturers expect you to know how to reference in their faculty's style, even if you have never referenced before. This leaves you relying on your tertiary institution's big 'How to reference' instructions or a library guide. This is probably a very comprehensive resource, but its many details, and many examples, may make your head swim.

This book would like to avoid both of these problems by covering the basic strategies and reasons for referencing as early in your tertiary academic career as possible. So, Chapter 2 seems like a good place to start. Before we do, we need to establish why it is so important to reference, or cite your sources, at all.

Firstly, referencing helps us to avoid plagiarising by acknowledging the original author or source of an idea. Unless you are writing an

Note!

Remember what we said in Chapter 1 about tertiary studies often not seeming fair? About your lecturers not babysitting you? I present to you Exhibit A: this 'unreasonable' expectation to reference without being taught how to might not seem fair, but it is what it is, and you need to figure it out.

Plagiarism is the act of using, copying, or closely imitating another author's work or ideas without his/her due acknowledgement or permission, or presenting another author's ideas or work as your own (Dictionary.com, 2020). For example, 'My lecturer accused Refiloe of plagiarising because she did not cite any of her sources in her assignment.'

opinion piece or reflective essay, just about every argument you make should use a reference if you want to avoid the academic 'naughty corner'. For example, using psychoanalysis to support a response in a psychology assignment without giving Freud the credit is pretty much stealing his ideas. By simply referencing him, you make your lecturer happy, and also strengthen your argument. This brings us to the next point...

Referencing is an excellent way to show that you have used credible or authoritative sources to support your arguments. Using Wikipedia as a reference is never advised, as it does not count as an authority, but referencing experts in your area is a good way to earn the trust of your reader (who is likely to be your lecturer). You should also benefit by being able to make better arguments, thus getting better marks.

Thirdly, when you reference in a way that shows you are familiar with different authors' views on a subject, you show your reader that you have read widely and engaged suitably with your topic. This does not mean that you need twenty different references for a single paper, but rather that you should aim for a variety of perspectives to present your case. This approach definitely has a positive effect on your perceived credibility too.

Finally, and perhaps most obviously, when you provide a reference for your source, you help readers of your paper to find the original sources. This allows readers to check information, or to use your sources to find related ideas that you have not included in your paper, but which might be of use to them. For example, if you cited Nick Middleton (2008) in a paper on soil erosion in southern Africa, your reader could refer to the source and discover valuable information on a topic that they need to research, such as tropical deforestation.

IMPORTANT

Please consult your tertiary institute's referencing policy. This may be referred to by a number of titles, including the Plagiarism Policy, the Academic Integrity Policy, or the Intellectual Ethics Policy, depending on your institution. This policy will outline the academic ethics your institution supports and desires, as well as what happens if someone infringes these. Because you do not wish to do any infringing, ensure that you are familiar with the specific rules and style guides of your faculty and institution before you submit any assignments.

 Use Student exercise LZ 2.6 on the Learning Zone to check your understanding of why it is important to reference and what plagiarism is.

2.5 Methods for referencing

In practice, we use two main methods of referencing sources:

- in-text referencing, which usually includes the author's surname and a year of publication next to the relevant sentence; and

- bibliographic referencing, which lists the full details of the consulted sources at the end of a paper, chapter or book.

It is important that you use both of these, and that they 'speak' to each other – for example, when you have mentioned Middleton (2008) in an in-text reference in paragraph 3 of your paper, the full details of the text should appear in your bibliography.

Before we show you how to reference, we need to consider which referencing style you will use. As we mentioned in the introduction to this chapter, different faculties prefer different styles of referencing. For this reason, you will find practical guides or links on the Learning Zone to four of the most likely referencing style options:

 See the Learning Zone for reference guide examples and links for some of the main styles of referencing used in southern African universities (LZ downloadable resources).

- the Harvard method, frequently used by business, humanities and social science faculties;
- the American Psychological Association (APA) style, a popular choice of psychology departments;
- the Vancouver style, which many science and health science faculties use; and
- the South African Journal of Human Rights' style, which is used by many law faculties in southern Africa.

This list if not exhaustive – in fact, Gill (2009) argues that there are 3000 ways to cite sources! – but you should be able to work out what you need to do by looking at these examples. For now, remember that you need to establish what referencing style you are expected to use so that you can practise your skills in as focused a way possible.

This next section will use the Harvard method on a few practical, annotated examples. For detailed guidance on these methods, you will need to visit the Learning Zone!

2.5.1 In-text referencing

The way you place your in-text reference will depend on how you wish to structure your sentence. You could, for example, decide to:

- keep your author name/s as part of the sentence; or
- not use the author name/s as part of the sentence.

Let us look at a few examples so you can see the different possibilities.

Let us assume that you are writing an education essay on the use of statistics in education. You have found some useful information in a book called *Statistics for the social sciences using Excel: A first course for South African students*, by Glyn Davis, Branko Pecar, Leonard Santana, and Alban Burke, which was published in 2014 by Oxford University Press.

In the paragraph you plan to write, you will briefly outline the concepts of mean, median and mode (don't worry if you don't know what these mean yet). Let's take one of these concepts, the mean, and give you a few simple examples of how you could do this:

> **INCLUDING THE AUTHORS IN THE SENTENCE**
>
> In statistical terms, Davis et al. (2014) describe the mean as an average, central value in a set of data.

You may have noticed a few things about how we did this:

- we did not include all the authors' names in the sentence – if your source has one or two authors, the name or names should both appear, but if there are more than two, list them all the first time you cite them, and thereafter use the Latin abbreviation 'et al.' (short for 'et alii'), which means 'and others';
- the year of publication appears in brackets, immediately after the authors' names;
- the sentence still needs to read like a complete, correctly punctuated sentence – try covering up the year of publication in the brackets to check this.

Note!

If you use 'et al.', please remember to put a fullstop (.) after 'al' and ensure that your sentence concord, or verb agreement, is appropriate for a plural subject.

You might find the following word combinations useful if you are trying to include your author names in your sentences and are not sure where to start (in these examples, X stands for the name of the author):

- According to X (date), …
- As X (date) establishes in his/her …
- X (date) argues that …
- X (date) takes the position that …
- Studies by X (date) suggest that …
- Pointing out that …; X (date) rejects the …
- In her study on …., X (date) theorises that …

> **EXCLUDING THE AUTHORS FROM THE SENTENCE**
>
> In statistical terms, the mean refers to an average, central value in a set of data (Davis et al., 2014).

Did you see what we did?

- The meaning of the sentences is the same, whether we include or exclude the author names, but in the latter, we restructured the sentence to move the authors out. It is sometimes easier to phrase your sentences without using the author names – just remember that sentence variety is better, so vary your in-text reference – sometimes including names, sometimes excluding them.
- If you choose to exclude the author names, they will appear as the first items after the open bracket, followed by a comma and the year

of publication. If you have quoted directly from a book, you will also include a page number – for the example we used, you would have said (Davis et al., 2014: 69) if you quoted a phrase from page 69.

These are just two, easy examples to guide your understanding of the principles behind in-text references; there are different ways of doing in-text referencing and sometimes you will need to cite multiple sources to support a single point. Examples of these more complex in-text referencing types may be found in your Learning Zone examples, so do consult that and the guide for your faculty.

 Use Student exercise LZ 2.7 on the Learning Zone to practise in-text referencing.

2.5.2 Bibliographic referencing

Whether your list of sources is referred to as a **reference list** or as a **bibliography** will probably depend on which referencing style your faculty has chosen. Some scholars (see Litting, 2013) take the position that a reference list includes *only those sources that you have specifically referred to in the body of your writing*, while bibliographies include additional readings that you did not specifically refer to. However, these meanings are not universally accepted, and the two terms are often used interchangeably.

Most lecturers would agree with Litting (2013) who advises that it is generally better to exclude references that you have not specifically referred to in your writing. It is therefore useful to check exactly which option your lecturer expects before you submit any assignments.

For every assignment, you ideally need to develop a reference list, providing details so that readers can locate the original texts for all of your cited sources. Although more complicated examples are provided in the Learning Zone guide, we would like to explain the process of creating a reference list here so that you have a solid understanding of the principles behind it. Let's look at an example of a Harvard bibliographic reference, sampled from the reference list for this chapter, to demonstrate some of these points:

SAMPLE REFERENCE LIST ITEMS

Du Toit, P., Heese, M. and Orr, M. 2000. *Practical guide to reading, thinking and writing skills*. Cape Town: Oxford University Press.

MacLeod, M. n.d. 'Types of reading'. *University of Calgary*. Retrieved 27/08/2019, available at <http://fis.ucalgary.ca/Brian/611/readingtype.html>

Roberts, P. 1996. 'Critical literacy, breadth of perspective and universities: Applying insights from Freire'. *Studies in higher education*. 21(2): 149–163.

Right, let's see what patterns we can detect in these examples using the Harvard style:

- Firstly, reference lists are compiled in alphabetical order, according to first surname – in this sample, surnames start with D, M, and R. If two or more authors have surnames that start with the same name, simply look at the next letter and keep to the alphabetical order. In some cases, the same author has more than one publication – in these cases, you would list the most recent publication first.
- The author initials (not the full name) come next, each one with a full stop. For more than one author, place a comma or 'and' after the fullstop, before including the next author's surname, then initials.
- The year of publication follows the author – some Harvard styles include brackets, while others use fullstops. *Whichever style you need to use, ensure that you follow it consistently.* You may note that under the *MacLeod* reference in the sample above, an 'n.d.' appears. This means that '**no d**ate' of publication can be identified for the source. It is not ideal as the reference is technically incomplete, but this is a relatively common problem.
- The title of a book is usually presented in italics after the date and is followed by a full stop. Journal and news articles are not italicised and may be shown *with or without* single quotation marks (just be consistent in your choice) – the title of the journal or the newspaper is italicised instead, and follows after the title.
- The place of publication is given; this is usually the city where the company's head office is. It is placed after the book title or website name, and is followed by a colon (:).
- The publishing company is placed last, after the place of publication.
- References to websites, such as the MacLeod reference in the sample, should include an indication of what site the article was sourced from, and should also show the date that you *retrieved* the article from the site, and the specific URL (web address), from which you accessed your source.
- You will note that the final source in the sample list refers to a journal, resulting in it needing a few unusual features in comparison to the book and website already discussed. In this example, additional information includes the volume (21), the issue number (2), and the page numbers (pp. 149–164) on which the article appeared.

Note!

Identifying the retrieval date is important for digital resources, since links are often updated, changed, or disabled and there is no guarantee of your reader accessing the same content you used.

There are many different types of sources, and many referencing styles, but the best advice may be to format your references consistently. Most lecturers understand that there are multiple ways of referencing, even within one style like the Harvard style. They therefore usually accept some small deviations from 'their' chosen style as long as the style that you have used has been *consistently* applied.

Please remember this when you are working on your assignments, and practise your referencing skills as often as possible using the Learning Zone activities for this section.

 Please visit the Learning Zone to complete the comprehensive referencing activities for this chapter. These activities relate specifically to the four main referencing style guides provided in the Learning Zone and referred to earlier. Please select the style guide most suitable for your purposes, and then complete the relevant activities for that style guide. The online assessment Chapter 2 Questions will also help to test what you have learnt.

Before you move on to the next chapter, revisit the objectives for this chapter, and draw an honest conclusion about your ability to perform or answer these fully. Ensure that you revise any elements that you do not feel comfortable with. If you are still struggling with any of the reading or referencing techniques, do not panic – all these skills need a substantial amount of practice to perfect. As long as you are honest about the limits to your mastery of them, and keep practising, you should be well on your way to improving your study skills as a whole.

 Use Student exercise LZ 2.8 on the Learning Zone to practise making a reference list or bibliography. Your lecturer may have a further activity for you to do in small groups.

2.6 Conclusion

This chapter introduced you to two essential academic skill sets that you need in your academic arsenal, namely reading and referencing techniques. It gave reasons for using a variety of solid reading skills, as well as why referencing is so important. Practical techniques and activities were offered to practise your reading, and general principles of both in-text and bibliographic referencing were outlined to launch your understanding of how to engage in the various referencing activities available on the Learning Zone.

References

Davis, G., Pecar, B., Santana, L. and Burke, A. 2014. *Statistics for the social sciences using Excel ®: A first course for South African students*. Cape Town: Oxford University Press.

Dictionary.com. 2020. *Dictionary.com*. Retrieved 05/02/2020, available at <https://www.dictionary.com/>

Du Toit, P., Heese, M. and Orr, M. 2000. *Practical guide to reading, thinking and writing skills*. Cape Town: Oxford University Press.

English, J. (Ed.) 2012. *Professional communication: Deliver effective written, spoken and visual messages*. 3rd ed. Cape Town: Juta.

Freire, P. R. N. 1970. *Pedagogy of the oppressed*. Translated by M. Ramos. London: Penguin Education.

Freire, P. R. N. 1985. 'Reading the world and reading the word: An interview with Paulo Freire'. *Language arts* 62 (1): 15-21

Gill, A. 2009. 'There are 3000 ways to cite source material – why not make it one?' *Times Higher Education News*, 25 June 2009. Retrieved 27/08/2019, available at <https://www.timeshighereducation.com/news/there-are-3000-ways-to-cite-source-material-why-not-make-it-one/407112.article>

Great Schools Partnership. 2015. In S. Abbott (Ed.). *The glossary of education reform*. Retrieved 27/08/2019, available at <http://edglossary.org/hidden-curriculum>

Hatzi, K. J. 2009. 'Making predictions: Keys to unlocking your textbooks'. *Slideshare*, 13 September 2009. Retrieved 11/01/2015, available at <https://www.slideshare.net/kjhatzi/textbook-structures>

Kruger, H. C. 2004. *Parameters for the tertiary training of subtitlers in South Africa: Integrating theory and practice*. Doctoral Thesis. North-West University.

Leavitt, M. n.d. 'Learning from visuals: How visuals can help students learn'. White Paper: *Wiley Visualizing Series*. Retrieved 31/01/2020, available at <https://www.wiley.com/college/visualizing/huffman/slides/Visualizing_White_Paper.pdf>

Litting, D. 2013. 'Difference between reference list and bibliography'. *UTS Library Answers*, 3 April 2013. Retrieved 27/08/2019, available at <http://www.lib.uts.edu.au/question/29624/difference-between-reference-list-and-bibliography>

MacLeod, M. n.d. 'Types of reading'. *University of Calgary*. Retrieved 27/08/2019, available at <http://fis.ucalgary.ca/Brian/611/readingtype.html>

Middleton, N. 2008. *The global casino: An introduction to environmental issues*. 4th ed. Kent: Hodder Education.

Oxford University Press. 2020. *Oxford learner's dictionaries*. Retrieved 05/02/2020, available at <https://www.oxfordlearnersdictionaries.com/>

Roberts, P. 1996. 'Critical literacy, breadth of perspective and universities: Applying insights from Freire'. *Studies in higher education*, 21(2): 149–163.

Wendell Holmes, O. 2013. *The autocrat of the breakfast table*. Transcribed from the 1873 James R. Osgood and Co. edition by David Price. Salt Lake City, UT: Project Gutenberg

Winter, C. 2019. '10 benefits of reading: Why you should read every day'. *Lifehack*, 4 June 2019. Retrieved 27/08/2019, available at <http://www.lifehack.org/articles/lifestyle/10-benefits-reading-why-you-should-read-everyday.html>

CHAPTER 3

Planning your assignments

'If we begin with certainties, we shall end in doubts; but if we will begin with doubts, and are patient in them, we shall end in certainties.'
– Sir Francis Bacon (1605: Book I:161)

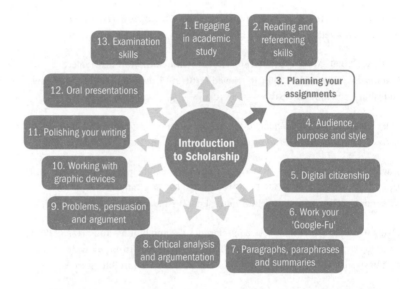

13. Examination skills

1. Engaging in academic study

2. Reading and referencing skills

12. Oral presentations

Introduction to Scholarship

3. Planning your assignments

11. Polishing your writing

4. Audience, purpose and style

10. Working with graphic devices

5. Digital citizenship

9. Problems, persuasion and argument

6. Work your 'Google-Fu'

8. Critical analysis and argumentation

7. Paragraphs, paraphrases and summaries

OBJECTIVES

At the end of this chapter, you should be able to plan academic essays with confidence by:

- understanding the value of using a big-picture approach to provide necessary context;
- correctly analysing questions using a topic analysis technique;
- identifying the nature and features of different types of academic responses;
- identifying the gaps in your topic knowledge;
- demonstrating basic online and database research skills to source relevant information;
- developing a logical essay framework;
- identifying various grammatical and syntactic elements necessary for effective persuasive writing;
- recognising the importance of revising and editing assignments prior to submission.

> **IRL**
>
> Francis and Lwazi's class has been given a particularly 'evil' first assignment for their Business Management 101 module. Although both acknowledge the difficulty of the task, they are dealing with it in very different ways: Francis is practising his usual 'ostrich in the sand' approach, and is determined not to worry about it until he absolutely has to, while Lwazi has decided to get his assignment done as quickly as possible. He has decided that even if he doesn't quite 'get it', he wants to tick it off his to-do list.

3.1 Introduction

Francis and Lwazi are certainly not unique in their approaches to challenging assignments – you may even recognise yourself in the way that one of them is approaching the assignment. Sadly, neither approach is likely to result in success: Francis runs the risk of not completing his assignment and will almost certainly do a rush-job, as will Lwazi, who may well miss the point of the task by simply aiming to tick it off.

Most importantly though, neither student is likely to learn from the challenging assignment, which will have been designed to give them an opportunity to explore their subject in more depth. Since one of the primary reasons you are studying further is probably to improve your understanding of the world, you can see that both Francis and Lwazi's approaches are far from ideal.

Okay, so if neither of these approaches is suitable, what is the right approach to assignment-writing? Unfortunately, there is no panacea that will magically make assignment-writing as simple as watching television.

A **panacea** refers to a universal remedy or cure-all type solution for something. For example, 'That politician argues as though democracy is a panacea for all the world's problems despite proof that it is not always the solution.'

The truth is that different approaches will work for different students, and what works for one type of assignment may not work for all assignments. For example, you may need to complete a practical project, an essay question, a work simulation, an online activity, or, very commonly, short questions or multiple choice questions. However, we can focus on and recommend one particular approach to academic essays, since the majority of your academic assignments will take the form of essays, and the principles for completing most of these are similar across different types of assignments.

This chapter will introduce you to an easy-to-follow six-step process, and provide you with an overview that will help you to plan for most kinds of assignment questions. The chapters that follow this one will then go into detail, providing you with opportunities to practise your skills at each of the six stages.

If you are confused about why this book uses a 'big-picture-then-detail' approach, try to imagine that you are standing looking up at the stars and planets at night – you may feel small, but you are likely to have a better idea of where you are in relation to the world than if you had not looked up at all. Taking this perspective helps to move your

Egocentric, in this context, simply means that you focus your attention on the world from your own perspective, and do not consider other perspectives, attitudes, beliefs, or ideas etc. For example, 'The commonly held belief that all the species on Earth were put there to serve human needs is another sign of human egocentrism.'

focus from an egocentric one to a more relative one, and helps you to consider different perspectives and possibilities. This view also assists you in staying focused on your end-goal, and allows you to adjust your approach to the details if necessary. Focussing on the 'big picture' is therefore important, and will form the focus of this chapter.

The chapters that follow will change your perspective again, asking you to examine and practise skills in a lot more detail. To continue our earlier example, in these later chapters you will be *looking in* at the detail of a chosen star or planet (instead of standing looking outwards at all the stars) and so be in a better position to relate your chosen star back to the universe you saw when you were having your 'big picture' moment.

If we apply this to your studies, having a solid idea of the big picture and then spending time on the details will help you to integrate all of those little bits of important information into your bigger-picture understanding of the assignment, as well as of your subject as a whole.

> If you take this even further and try to understand the 'big-picture' goals of your chosen degree, its role in society, or even of your own long-term role in society, you will also more easily make connections between other subjects, topics, ideas and knowledge. This should enhance your academic success as well as your potential contribution to society as a productive and adaptable citizen, so there is even more reason you should look at the 'big picture' more frequently!

For now, let's keep our aspirations reasonable and take a look at our 'big-picture' view of approaching assignments...

Figure 3.1. A six-step approach to assignments

1. Do your topic analysis	Analyse the verb/s and specific instructions in the question
2. Mind the (knowledge) gap	Identify what you already know (or think you know) about the topic, identify your thesis point, and what you need to find out to answer the question fully
3. Work your Google-Fu	Develop focused search phrases to find relevant information online, in prescribed or printed books, and in journal databases
4. Build your skeleton	Organise the relevant information you have sourced into a logical, paragraph-by-paragraph skeleton that builds your argument convincingly
5. Draft your essay	Flesh out your skeleton by writing the first draft
6. Revise – edit – revise again	Revise your first draft by checking your thesis statement, topic sentence, and supporting sentences. Edit for language and revise again before submitting

3.2 A six-step approach to assignments

Think back to your high school years: did you ever receive an assignment back from a teacher that resembled a bloody battle of red pen? Perhaps accompanied by comments like 'rambling', 'woolly', 'language!', 'paragraphing', 'irrelevant', 'unclear', or worse, simply 'No!'?

If we are honest, most of us have received feedback of this kind, even if we did really well at school. Hopefully, if you did get this type of feedback, it made you determined to improve your efforts, and focus on planning your work better. If it did not inspire you to improve then, the bad news is that at the tertiary level, you simply cannot afford to submit unstructured, sloppy, unedited, or irrelevant work. It is time to raise that game of yours!

The six-step method shown in Figure 3.1 above, and discussed in more detail in the later chapters of this book, should help to make your lecturer's feedback include words like 'cogent', 'lucid', and 'well-argued'. Let's aim for that and take a look at the process below.

As you can see from Figure 3.1 above, the process is not complex or difficult. However, it does require that you are really thorough, so make sure that you keep all six of the steps in mind as you work your way through this book, and also as you complete your assignments. The rest of this chapter will explain each of these steps to develop your 'big-picture' understanding of the process, and will use the following generic assignment question to provide practical examples along the way:

In an essay of between 800 and 1 000 words, examine the role that socio-economic factors play in hindering voter participation in South African national elections.

3.3 Topic analysis

The first step in this six-step process is to figure out exactly what your lecturer is expecting from you. Socrates rightly argues that understanding a question is half an answer (Plato, 2005), and yet, strangely, few students pay enough attention to this step, despite many having already received unfortunate marks as a result of understanding a question poorly. Neglecting this stage could have huge and negative consequences for your assignment mark, but it also means that you have wasted your time researching and writing about something irrelevant – and wasted your lecturer's time in marking your sad and off-topic attempt.

To properly execute this first step, you need to perform a thorough topic analysis to establish:

- the format you are being asked to use;
- the topic of the assignment;
- the specific instructions about the extent and depth to which you need to respond.

Biographical note

Socrates was a classical Greek philosopher who lived between 470–399 BCE, and who contributed significantly to the fields of ethics and education.

"Try to think of each sentence as a tweet sent by a celebrity named William Shakespeare."

To conduct a thorough topic analysis, we will make use of a topic table – this is something simple that you can create using a sheet of paper, a ruler and pen, or by using word processing software capable of creating tables. Your blank table should look roughly like the one in Table 3.1 below.

Table 3.1. Basic outline of a topic analysis table

Stage 1: Identify required format		
Stage 2: Identify main topic	Stage 3: Identify each instruction word (verbs)	Stage 4: Identify delimiting words (for each verb)

3.3.1 Types of academic responses and formats

Before we start analysing the specific elements of the question, it is important for us to know what format we need to use to respond to the question. For example, there is a considerable difference between a multiple choice questionnaire and a question that requires an exhaustively researched essay; knowing which type to use will help you to respond appropriately.

Table 3.2 below looks briefly at some of the most common types and features of academic responses that you may have to write. Please remember that there are many other possible responses – these are just the most common ones you may encounter.

Table 3.2. Types and features of common academic responses and formats

Type	What it does	Common features
Multiple choice questions (MCQ)	*Efficiently measure the level of knowledge or thinking skills that you have on a topic*	MCQs are often poorly framed, but when done *well*, an MCQ assignment: · can examine content in broad, and even deep ways; · uses a posed question, with multiple answers from which you need to select to respond appropriately; · may provide more than one answer that is technically correct, so your recognition and discrimination skills are also tested; · is frequently used in undergraduate studies, particularly in the first years of study; · is also often used in economics and physics, where extra calculations will need to be done by students before selecting an answer.
Paragraph response	*Evaluates your understanding and ability to apply smaller bits of theory or concepts*	Depending on the marks allocated, a paragraph response should include: · a topic sentence that answers the question; · at least one example or reference that supports the first sentence; · an explanation of how your example or reference answers the question; · a concluding sentence that ties up the paragraph.
Analytical/ Literary essay	*Presents your analysis of a literary work such as a poem, book, or play*	· An introduction that presents the text and main issue/thesis under discussion; · the body of the essay will include analysis of the text itself in terms of stylistic and literary devices, supported by valid examples etc. in relation to the greater issue; · throughout the analysis, personal, critical and supported commentary that shows deeper understanding of the work's shortcomings and achievements in relation to the issue is included.

Table 3.2. Continued

Type	What it does	Common features
Argumentative essay	*Presents a strong and persuasive position on a particular topic to educate and convince the reader of your point of view*	• Strong views are presented in a thesis statement, in an often controversial manner to an audience not inclined to agree; • usually, both sides of an argument are presented, but one side's arguments are methodically dismissed; • may take the form of a point-by-point argument and refutation, or may present all arguments in support in one place, and follow them with arguments against; • final paragraph clearly states the conclusion that the author wants the reader to accept.
Discursive/ Expository essay	*Presents an objective examination of a topic and is one of the most commonly used forms of academic assignment and essay assessment*	• Presents different sides of an issue in as objective a manner as possible; • supported by facts and research; • the purpose is to present a balanced argument, not to convince the reader; • uses a formal and impersonal style; • does not have to be exclusively neutral; the author may draw (cautious) conclusions about the topic, or leave the reader to decide.

 The Learning Zone has examples of most of these types of essay so you can get a real feel for how they differ: see the Table 3.2 extension there.

Let's see how **Stage 1** works in practice.

First, you need to read through the hypothetical assignment question provided in Section 3.2 again *carefully*. Got it? Excellent – read it once more, please. Good.

Now, if you examine the different types of academic responses in Table 3.2 in relation to the earlier assignment question, you should note two important points:

- that the question specifically tells us to write an essay (without telling us what type) of between 800 to 1 000 words; and
- that out of the three possible essays in this table, a **discursive essay** looks most promising as this assignment has asked us to look at socio-economic factors that affect voting participation – the question does not really deal with a controversial issue, and is clearly looking for a more research-driven response than an emotional or persuasive one. As a result, we can use a discursive, and more formal, objective approach to the topic.

So, we have essentially completed Stage 1 of your topic analysis for this question. All we have to do now is fill in our conclusions about what format we need to respond in (see Table 3.3 below). Remember that you do not need to write out your notes in full sentences – the table is there to help *you* to break down exactly what *you* need to do, so shorthand notes are perfectly fine!

Table 3.3. Topic analysis – Stage 1

Stage 1: Identify required format	Discursive essay 800 – 1 000 words	
Stage 2: Identify main topic	**Stage 3: Identify each instruction word (verbs)**	**Stage 4: Identify delimiting words (for each verb)**

Once you have a perspective on the format you need to use, **Stage 2** requires that you identify the **main**, or general, theme of the topic. Keeping this in mind helps you to place the question in context, and allows you consider how the question ties in with your subject on a larger scale.

To do this, read the question through carefully again, and ask yourself how the question as a whole fits in to the objectives of the course you are studying. At this stage, do not try to limit the scope or topic at all – read through the question and try to summarise what you think is the **overarching theme** of the essay in no more than five words, as indicated in Table 3.4 below.

Table 3.4. Topic analysis – Stage 2

Stage 1: Identify required format	Discursive essay 800 – 1 000 words	
Stage 2: Identify main topic	**Stage 3: Identify each instruction word (verbs)**	**Stage 4: Identify delimiting words (for each verb)**
South African elections		

Clearly, it will not do to leave things there – if you simply started writing an assignment about South African elections, you would not come close to answering the question appropriately.

We need to move to **Stage 3**, and identify all of **the instruction words/ verbs** in the question. These words or phrases tell you what you need to do. Remember that many questions have more than one verb or verb instruction phrase, so it is really important that you get this step right and find *all* of the relevant instructions. Let's have a look at our example again:

In an essay of between 800 and 1 000 words, (examine) the role that socio-economic factors play in hindering voter participation in South African national elections.

Although this particular question instructs you to write an essay, this first part of the instruction does not actually form part of the election-related question itself. We have also already noted this requirement, so for now, let's focus on the only remaining verb in the question, 'examine'. Once you have identified all relevant instructions, you simply need to write them into your table, as shown in Table 3.5 below.

Table 3.5. Topic analysis – Stage 3

Stage 1: Identify required format	*Discursive essay* *800 – 1000 words*	
Stage 2: **Identify main topic**	**Stage 3:** **Identify each instruction word (verbs)**	**Stage 4:** **Identify delimiting words (for each verb)**
South African elections	*examine*	

Before we run gung-ho to the final stage of our topic analysis, let's consider what this particular instruction word means. To 'examine' a topic means that you are expected to present a point of view, give the main reasons for and against a topic, and come to some form of *supported* conclusion. As a result, you will need to do research to find credible sources, and then describe and interpret the relevant information you have found. Keep your instruction words in mind while you are busy planning and collecting suitable evidence for your essay – it will help you to source relevant information and deal with it appropriately.

Moving away from our example for a moment, it is important that you are able to interpret what different instruction words or phrases require of you, since you will have to deal with a variety of them in the course of your studies. These words or phrases are usually explained in detail in your particular institution's library reference guide, or in your study guides, so please make sure that you refer to the relevant document when you are analysing your topic. For now, a short summary of some common instruction words is included below to introduce you to the different requirements of each.

A more extensive list of instruction words is available on the Learning Zone for your uᴄᴄ. You can also use student exercise LZ 3.1 on the Learning Zone to practise working with argumentative essays.

3.3.2 Common instruction words

analyse	Break up the subject into its main ideas to uncover its nature, function, proportions, or relationships etc. – remember to support your breakdown with evidence.
argue	Put forward a point of view in a structured, logical manner – you may present both sides of an argument, but should clearly argue in favour of one side.
compare	Look closely at two or more things to examine their similarities and differences in an analytical/methodical way. (Don't get confused with the instruction to 'contrast', which looks only at differences).
critique/critically analyse	Examine the positive and negative points of a given topic/theory through reasoned argument and personal judgement based on supported evidence. You will need to demonstrate significant insight for this type of question.
define	Give the precise, concise and authoritative meaning for a term or concept in a way that distinguishes it from closely related subjects; where necessary, provide an example to illustrate the term.
demonstrate	Provide a step-by-step procedure to show how something is possible.
describe	Write a detailed account, in your own words, of how, what, or why something happens.
discuss/ examine	Present a point of view, give the main reasons for and against a topic, and come to some form of supported conclusion. You will need to describe and interpret the content for this type of question.
evaluate	Estimate the value, relevance or worth of something in a methodical way, and support your personal opinions with reasoning and/or evidence.
explain	Analytically outline how or why something happens and provide relevant examples of its occurrence. Look for cause and effect and remember to think critically, beyond the obvious, factual elements.
illustrate	Provide specific examples to clarify an explanation. These examples can take different forms, including words, and do not need to be drawn!
justify	Give a reason or reasons for your support of a particular argument or point of view.
outline	Provide a brief and organised overview, description or explanation of a particular topic – exclude minor detail here.
prove	Confirm or verify a statement by using formulae, or by conducting or citing experiments, tests, or logical arguments.
suggest	Provide or propose (supported) solutions, reasons or ideas in response to a particular issue.
summarise	Convey the main points of a text in as concise a way as possible using your own words, and without injecting your own content into the text.

Au fait is an adjective that came into English from French, so it is pronounced 'oh-**fay**' with a silent *t*. In English, it is commonly used to describe a person with a good knowledge of something. For example, 'Susan is completely au fait with the rules of the game.'

To **delimit** something means to set the limits or boundaries of something. For example, 'In South Africa, rivers, mountains and oceans delimit our borders'. In this particular case, it means to set the scope of your topic by identifying the key ideas.

Okay, now that you are a little more au fait with academic instruction words, let us move to Stage 4, the final step of our topic analysis.

For **Stage 4**, it is important that you identify all of the necessary delimiting words and phrases so that you clearly identify the full scope of the topic.

In Stage 3, we demarcated the main topic as relating to South African elections, but we noted that our essay would not be adequately written if we simply wrote on this topic – we need to refine it and determine what *specific elements* of South African elections we are expected to write about. Okay, let's see how Stage 4 works in practice:

- examine the question again, and highlight all of the main ideas/phrases;
- move these into your table, and then focus your *key ideas* by underlining them;
- keep the relevant words (such as 'play' in the table so that you understand the direction to take with the key words), but don't underline them, as indicated in Table 3.6 below.

You might have selected or grouped your words slightly differently to the example in Table 3.6, but what is important is that you identify *all* of the necessary words that lead you to the correct topic. This step is essential as it will guide the development of your search phrase if you are doing online or database research, and it will also help you to identify the relevant parameters for your essay topic.

Table 3.6. Topic analysis – Stage 4

Stage 1: Identify required format	Discursive essay 800 – 1000 words	
Stage 2: Identify main topic	Stage 3: Identify each instruction word (verbs)	Stage 4: Identify delimiting words (for each verb)
South African elections	examine	role socio-economic factors play hindering voter participation South African national elections

For this particular example, you may have identified that the core of what is required for this essay is examining the 'socio-economic factors' that 'hinder participation' in 'South African national elections'. Okay, but what exactly does this mean?

Well, it simply means that you need to find particular socio-economic factors relevant to the South African context, that reduce citizens' ability or willingness to vote in national elections.

To keep your essay tight and on-point then, this means that your essay will not look at other factors (such as political or cultural), rather focusing exclusively on socio-economic factors. You will also ignore socio-economic factors that influence voting in countries other than South Africa, or socio-economic factors that affect voting for provincial or local elections, as you have only been asked to examine the influence of these on *national* elections.

We hope this example has demonstrated the process of analysing your topic in a way that sparked multiple lightbulb moments for you. If you are unsure of how we got to this point, re-read the various stages in this section to see if close reading makes things any clearer. Remember that one of the best ways to improve your understanding is by practising as much as possible, so try this if you are still unsure.

Practise using a topic table on various extended writing or essay-type questions from different course assignments. This will help you to master using the technique, as well as helping you to prepare for your assignments!

By now, you should have a good idea of what your lecturer is expecting from you. The next step is to figure out what you already know about the subject, and what you still need to find out.

 Use the student exercise LZ 3.2 on the Learning Zone to work through a guided practice using a topic table.

3.4 Mind the (knowledge) gap

As you were working through your topic analysis, was your brain making connections with what you already knew about the topic? Were you able to identify what all the concepts meant?

Unfortunately, the topic given in the example in Section 3.2 is likely to be unfamiliar to you (unless you have some basic political science under your belt). For example, you may have no idea what a 'socio-economic factor' is. If you have never voted in a South African election, you may not have personal experience of voting and so may be uncertain of what it entails. You may also have no idea what the word 'hinder' means!

Despite that, you may indeed be familiar with the concepts of voting and national elections, and may have some personal experience either of voting, or deciding not to vote. So, as with most topics that you analyse, you will have some idea about the general context that you will launch your research from.

A quick way to establish what you know, and what you still need to learn, about your topics is to follow this process:

1. On a blank piece of paper, brainstorm everything that you know, or think you know about both the main topic and its specific, delimited areas. Do not hold back or try to edit yourself at this point – let your hand scribble notes as you think about them, even if what emerges is unrelated to the topic. If you cannot think of anything, or do not understand a particular element of the topic, simply write what comes to mind: even '*what????*' is

A **mind map** is a diagram that is used to organise information visually, and consists of a central concept, surrounded by a number of arms, each of which may, in turn, have its own arms. For example:

perfectly acceptable. Leave your ideas to settle for a while before proceeding with the next stage.

2. Now look at the product of your brainstorm. Using four different coloured highlighters:
    > highlight in green the areas that you were able to write about from personal experience;
    > highlight in blue the areas that you had specific academic knowledge of;
    > highlight in orange the areas that you had no idea about, or that you were unsure of; and
    > highlight in pink the areas that you can immediately identify as irrelevant.

3. On a new sheet of paper, use a **mind map** as below or table to summarise your brainstorm notes in a more organised way. Your new note should clearly show all the necessary elements of the essay, as well as reflecting the relevant <u>green</u>, <u>blue</u> and <u>orange</u> notes that apply to each of these elements. Exclude the irrelevant information highlighted in pink.

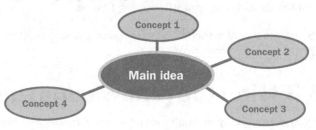

4. The result should be a clear indication of the gaps in your knowledge, and of the specific elements you will need to research.

Now that you are armed with a solid understanding of what your essay requires, and the information you need to find, you are ready to proceed to Step 3.

Don't forget to visit the Learning Zone — student exercise LZ 3.3 there will give you the opportunity to hone your brainstorming skills.

Note!

You will need to research most, if not all, the elements of your essay, even if you have relevant personal experience or knowledge. This is because your experience does not necessarily reflect everyone else's, and your knowledge might be outdated, or perhaps correct but need support in the form of evidence. It may even be incorrect! In an essay that is looking for formal, objective and credible information, your personal knowledge and experience will not elicit joyful acclaim from your lecturer. So, value your experiences and knowledge – they may help to shape your argument – but remember to verify and support them with researched information.

3.5 Work your Google-Fu

The third and critical step in your assignment preparation is to engage in gathering information. Your assignment, regardless of how well you are able to write, relies on you having found the most relevant and up-to-date information possible. If your sources are outdated or unreliable, your entire argument could fall flat.

Although this section's heading refers to working your 'Google-Fu', note that the phrase *Google-Fu* has become informally synonymous for the ability to find answers to questions using the Internet, *regardless* of the search engine or database you use. We will discuss the pros and cons of this in Chapter 5; for now, if you tell someone that their Google-Fu is strong, it means that they are able to find relevant, useful information in a skilful way.

Even if you have no 'Fu' at all, don't worry too much right now. You might be surprised at how little skill even frequent users of computers, social media, and the Internet have in finding relevant, credible academic information – your skills in this regard will be developed shortly.

During this phase of your assignment preparation, you should maximise your chances of sourcing relevant and credible sources by completing the following process:

- Revisit your final topic analysis table from Stage 1, and your final notes from Stage 2 and reread everything.
- From these notes, establish the key phrases or words that describe your topic. Try to include a broad subject as well as your delimiting phrases or words that restrict the topic.
- Now start digging – use your keywords to search:
  > the indexes of prescribed or recommended books on the broader topic in your library – if you have any questions, ask your librarian for help with finding the most likely books for your search;

> **Point of interest**
>
> Google is actually both the name of a multinational corporation and its core search engine, which attempts to organise and access information on the World Wide Web.

> **Note!**
>
> Chapter 6 of this book will help you to develop your Google-Fu, progressing from simple white-belt searches, to more complex black-belt analyses using search engines and online databases. You will quickly realise how valuable a black-belt in Google-Fu can be!

> **Note!**
>
> In general, finding suitable sources at the undergrad level is easier than it sounds. For example, if we enter the words *socio-economic factors hinder participation South Africa national elections*, all of which we identified in Stage 1, even without adding any fancy Google-Fu ideas like 'search operators', the returns are easily relevant to our essay. For example, in first place pops up a survey on South African voter participation in elections, prepared by the Socio-Economic Surveys Unit of the Human Sciences Research Council; following that, a number of other potentially relevant academic articles. The first option is ultimately not ideal because it is from 2005, so a little too old for our purposes, but you are definitely likely to find relevant and credible information from some of the other returned items.

  > the online databases of journal collections in your campus library – again, your librarian should be very helpful if you find yourself struggling with the database;
  > the friendly 'Google' search engine itself.

- Use your skim-reading skills to determine whether or not the returned items, chapters, or points are relevant to your specific essay. As soon as you identify something potentially useful, make sure that you record its bibliographic detail on the back of your table or mind map from Stage 2, or copy it into a document on your computer. This will help you to find that source later, but also save you time when you are writing the reference list for the end of your assignment!
- When you sit down to read your identified sources, make a note of any important points, earth-shattering quotations (they've got to be brilliant to be quoted!), or concepts, as well as their locations next to the relevant keyword or phrase on your final note from Stage 2. This will help to keep your search organised, and allow you to note any remaining gaps or areas that you may have found too much information for.
- Repeat the search process until you think that you have sufficient information and/or sources to complete your assignment. *Hint*: Try using different keyword or phrase searches that specifically relate to the area of your essay for which you did not find enough information in the first round of searching.

Once you feel that you have enough information to build an argument that is relevant to the given question, you can move to Step 4 of your assignment preparation.

 The Learning Zone has a student exercise that will show you how to integrate and develop your search skills or 'Google-Fu' to great purpose. See LZ 3.4.

3.6 Build your skeleton

Take a look at the image below:

You might already be familiar with Auguste Rodin's sculpture of *The Thinker*, which currently exists in 28 giant bronze versions and a number of smaller plaster castings around the world. Rodin's 1881 sculpture is a critically acclaimed work of art, but few people stop to consider that an exceptionally strong framework is required in order to create something as impressive as this artwork; the framework may well need a very particular structure in order to support the heavy sculpture effectively.

This approach should be applied religiously to your essays too: if you are aiming for a masterpiece deserving of critical acclaim (and excellent marks), you need a solid framework, or skeleton, to support and structure your essay.

The fourth step in preparing your essay is to construct a rough, but logically arranged, framework that will act as the support for your essay. Since you have already completed Stage 2 and 3, and are already in possession of the main ideas, key points, and necessary information, this step will be a lot easier to accomplish.

But before you attempt to convert your annotated mind map or table from Stage 2 into a skeleton, we need to get through a quick anatomy lesson – that is, the anatomy of an essay. You will find that most essays follow a reasonably generic format and that they all consist of similar elements. These include:

- a title;
- an introduction:
  - > hooks your reader in an interesting way;
  - > includes your **thesis statement**, a sentence that clearly demonstrates the position you will take in the essay;
  - > introduces or signposts the order in which you plan to detail the points and evidence.
- body paragraphs (the number of these depends on how long your essay needs to be):
  - > each paragraph has *one* main idea, or **topic sentence**;
  - > each topic sentence is supported by reliable, relevant evidence in the form of **supporting sentences**;
  - > each paragraph uses a concluding sentence that links it to the next paragraph, and provides a solution, prediction, or answer.
- conclusion:
  - > signals the end of the essay;
  - > can be a brief summary or restatement of the introduction, a call to action, or a strong statement reaffirming how the essay has proven the thesis statement, depending on what kind of essay was written;
  - > avoids use of trite expressions like 'in conclusion'!

For Stage 4, you will need to develop a skeleton that meets these requirements. Start working in a Word document now if you have not already done so. This is because your essay will start to develop quite quickly from here, and working electronically allows for a more fluid and easy approach to shifting the order or sequence of your content than working on paper does.

The easiest way to write your skeleton is to copy the rough outline provided here, and then populate it with your notes from Stage 2 and

Stage 3. Keep all of your main points to the left of the page, and indent any supporting evidence immediately underneath them so that they are clearly grouped together. This will allow you to easily identify your thesis statement, topic sentences and supporting sentences, and also to identify, once you have populated your skeleton, if you have any 'ribs' missing!

Do not worry about writing in full sentences at this point. Instead, focus on getting your information sorted into a logical sequence, and ensuring that:

- you have covered all the required elements of the essay;
- your thesis statement is present;
- your topic sentences are all accompanied by sufficient supporting sentences.

Once you have a complete skeleton, take note of any gaps or irrelevant content. Source extra information if necessary, and be ruthless about cutting out information that does not speak directly to the task. You will save a lot of time if you resolve these issues at this stage. If you are satisfied that your skeleton will hold up to the 'Rodin test', you are ready to proceed to Step 5.

3.7 Write your draft

The golden rule for a first draft is to **write at speed, and revise at leisure**. However, if you are working from a well-crafted skeleton, your first draft should be much simpler to write since the basic plan and sequence is already decided. This is the point at which you start to add 'flesh' to your skeleton by filling in words, and connecting your bullet-type points into full, although not necessarily grammatically correct, sentences.

Chapters 7, 8, 9 and 11 will cover writing skills in more detail, and will specifically deal with how to structure thesis statements, topic sentences and supporting sentences. They will also try to hone your skills in writing persuasive paragraphs and paraphrases. Since these are important elements of writing essays, we would advise you to pay attention to these chapters when you are preparing an assessment response.

It is important that you should not worry about spelling, grammar or punctuation when you are writing your first draft. Rather try to achieve a cohesive flow to your argument, and leave the revision and editing for the final stages.

3.8 Revise – edit – revise

You made it to **Step 6**, the last in preparing your essay. Don't get too excited yet though – the revision, editing and repeat revisions of an essay can take many cycles, especially if you are determined to get good marks!

Although Chapter 11 of this book covers the elements of revision and editing in some detail, we will take a bare-minimum look at the three steps for this stage. These are:

- Revise your first draft:
  - > Revision means that you read through your *argument* critically and make any necessary adjustments to it – this is not the same as editing, so you still do not need to focus on grammar and punctuation etc.
  - > Check that your introduction clearly introduces your thesis statement and that it hooks the reader.
  - > Check that your argument follows a logical sequence, or rearrange it.
  - > Check that each paragraph has the necessary elements, including references, and that each paragraph contains only one main idea.
  - > Remove any irrelevant information that you missed in Step 4.
- Edit your essay:
  - > Okay, *now* you are ready to go all spellcheck and 'Grammar Nazi' on your work.
  - > Ensure that you have used full sentences, proper paragraphing, and that all of your work is correctly punctuated.
  - > Check that your in-text referencing has been correctly formatted and that it matches your reference list at the end of the essay.
  - > Check that your paragraphs connect naturally to each other by using suitable transitions.
- Revise your draft again:
  - > Repeat the steps from the first draft – after you have edited your work, your final revision should reveal a cogent, easy-to-read, and convincing argument. Do not feel scared to keep revising and moving content around if necessary – just remember that with each change, you will need to do another edit, and another final review to ensure that you have not left any sentences hanging, or made any new grammatical mistakes.

If you have made your way through all six stages of your essay preparation, you should feel confident that your essay will meet the expectations of your lecturer.

 You can check your skills on the Learning Zone by tackling student exercise LZ 3.5, which brings together the skills you have learnt in this chapter.

3.9 Conclusion

This chapter introduced you to a six-step process for approaching essay assignments. By using topic analysis, identifying your knowledge gaps, sourcing information, developing a skeleton, writing a draft, and then revising and editing it, you will have covered all of the vital elements involved in writing assignments that are built to impress.

References

Bacon, F. 1605. *The advancement of learning*, edited by Joseph Devey. New York: P. F. Collier and Son. Retrieved 03/02/2020, available at: <https://oll.libertyfund.org/titles/bacon-the-advancement-of-learning>

Plato, 2005. *Protagoras and Meno*. Translated by Adam Beresford. London: Penguin Classics.

CHAPTER 4

Audience, purpose and style

'Prose is architecture, not interior decoration'
– Ernest Hemingway (1932:191)

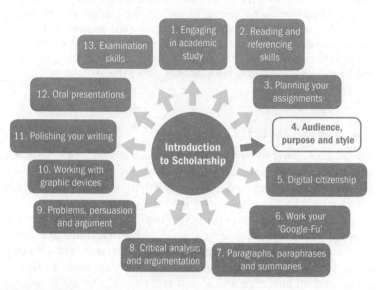

1. Engaging in academic study
2. Reading and referencing skills
3. Planning your assignments
4. Audience, purpose and style
5. Digital citizenship
6. Work your 'Google-Fu'
7. Paragraphs, paraphrases and summaries
8. Critical analysis and argumentation
9. Problems, persuasion and argument
10. Working with graphic devices
11. Polishing your writing
12. Oral presentations
13. Examination skills

Introduction to Scholarship

OBJECTIVES

At the end of this chapter, you should be able to:

- explain the importance of identifying your audience and purpose;
- identify the purpose of your written or oral task;
- analyse your audience using a given technique;
- understand the differences between objectivity and subjectivity, fact and theory, and opinion vs persuasion;
- recognise different levels of formality in writing;
- select and maintain a suitable style of writing for your task;
- employ a variety of devices to improve readability.

Roberto is a first-year law student, and is under a lot of pressure to perform well because of the sacrifices his mother and grandparents have made for him to study. Roberto knows that he did not 'shoot for the stars' at school, so he has asked one of his classmates, Tarachand, if she would give him some feedback on his first assignment before he submits it. He hopes this will improve his chances.

Tarachand agrees, knowing that this will help both her and Roberto, but after looking at Roberto's assignment, she is worried that her feedback will offend him – unfortunately, most of it is negative. Roberto has used very informal and emotional language, including slang, and has not provided credible, objective evidence. In addition to these issues, he has also wandered off topic to make unnecessary personal judgements, and Tarachand has no idea how to tell him all of this in a kind, constructive way.

4.1 Introduction

The word **commend**, used in this way, means to praise someone, usually in a more formal way.

In the IRL scenario above, Roberto should be commended for his attempt to improve his writing skills and his marks. Asking a classmate to review your work is an excellent idea – especially if that classmate understands the project and has done their own work diligently.

However, a very important part of asking for feedback is to be open towards the feedback given. In the scenario above, it is vital that Roberto does not take Tarachand's criticism personally, and that he can step back and examine his assignment as objectively as possible: *Has* he used emotional language and slang? *Did* he use unreliable sources for his argument? *Has* he deviated from the task by making personal judgements? If Roberto can examine his assignment and the feedback honestly and unemotionally, he has a much better chance of correcting any issues in his assignment and receiving a solid mark for his efforts.

Prose refers to written or spoken language in its usual grammatical form, without any particular poetic or rhythmic structure. It is like natural speech, rather than poetry or song. For example, 'The novelist Salman Rushdie is a superb writer of prose.'

This chapter will guide you through a number of important 'architectural' prose elements. As Hemingway suggests in the introductory quote for this chapter, creating a clear and solid framework for your prose will help you to prepare a firmer, more impressive argument – getting straight into the 'interior decorating' of your paragraph writing is a much more dangerous approach if you want to be taken seriously.

In this chapter, we will look at identifying – and sticking to – your purpose; identifying your audience; understanding the differences between objectivity and subjectivity; selecting and keeping to a particular level of formality in your writing; and finally, using a variety of devices to improve the readability of your work.

4.2 Identifying the purpose of your task

Contrary to what you may sometimes feel, lecturers do not give you assignments in order to ruin your social life. They also do not give you assignments because they are bored and long to spend hours covering your efforts in red pen. Lecturers also do not give you assignments so that you can generate incomprehensible, **vacuous**, academic gobble-degook for them to look at.

The word **vacuous** refers to something that is empty, and in this case, it refers to mindless assignments without substance.

"I SPENT *FIVE* HOURS WORKING ON MY REPORT! ONE HOUR TO GO TO THE MALL FOR AN INK CARTRIDGE, TWO HOURS ON HOLD WITH TECH SUPPORT, 45 MINUTES LOOKING FOR A SHEET OF WHITE PAPER, 30 MINUTES SEARCHING FOR THE PERFECT FONT..."

There are, however, a number of really good reasons that lecturers do give you formative assessments like assignments; for example, to:

- monitor whether or not learning is taking place so that they can adjust their teaching methods;
- provide critical feedback and guidance to you about your current learning or understanding;
- stimulate critical and creative thinking;
- prepare you for real-world events and problems;
- help you to develop desirable attitudes to study habits and academic practice.

Apart from these valuable but intangible purposes, there is also a **tangible** reason that lecturers give you assignments to complete: assignments allow your lecturer to grade your current level of mastery over the given work in the form of a mark, or percentage. In turn, this allows you to establish how much more work you need to do in future, or alternatively, lets you impress your parents and friends.

Tangible, as used here, refers to something that you can perceive through the senses, most notably the sense of touch. For example, 'Kelebogile used a number of tangible arguments to support her claim.' Intangible means something that cannot be sensed or touched.

The specific purpose of the assignment *you* have been given, however, may vary depending on the topic. For example, your lecturer may want you to prove something, to argue against or analyse a statement or topic, to combine different elements into a new idea, or even to present various sides of an argument around an issue. In short, your assignment should provide you with an idea of the level of **cognitive engagement** expected from you.

Different **cognitive strategies** and levels were outlined in Chapter 1 of this book. Revisit it for a quick refresher if necessary.

The previous chapter took you through the process of analysing your topic; this is an essential component of understanding the purpose of your assignment. Topic analysis tells you what your lecturer is expecting from you. This influences the type of information that you gather, as well as the approach that you use in presenting your topic.

Your topic analysis should therefore have helped you to decide on the purpose of your task. Ask yourself if you need to *convey facts*, to *convey opinions*, or to *persuade* your reader. Once you know this, you can decide on a suitable approach and style for your task. In general, the approach you take will depend on your purpose, as can be seen in Figure 4.1 below.

Figure 4.1. Purpose, approach and formality

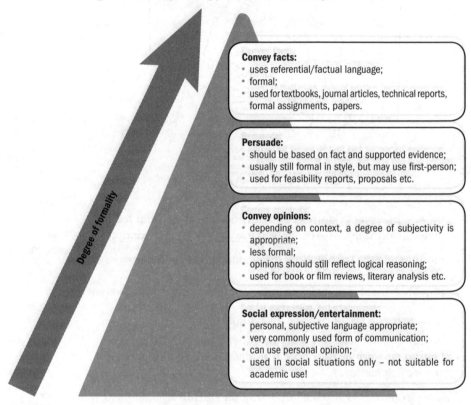

Convey facts:
- uses referential/factual language;
- formal;
- used for textbooks, journal articles, technical reports, formal assignments, papers.

Persuade:
- should be based on fact and supported evidence;
- usually still formal in style, but may use first-person;
- used for feasibility reports, proposals etc.

Convey opinions:
- depending on context, a degree of subjectivity is appropriate;
- less formal;
- opinions should still reflect logical reasoning;
- used for book or film reviews, literary analysis etc.

Social expression/entertainment:
- personal, subjective language appropriate;
- very commonly used form of communication;
- can use personal opinion;
- used in social situations only – not suitable for academic use!

Degree of formality

It should be clear from Figure 4.1 that more academic settings are likely to require you to regularly engage in the purposes at the higher end of the pyramid, although social expression, at the lower end of the pyramid, is very common and does not require academic rigour. This will mean that your approach in academic settings will be more formal than your usual social expression allows for. Sections 4.4 and 4.5 of this chapter will guide you on this.

With your topic analysis fresh in your mind as you prepare to turn your essay skeleton into something more complete, it would be useful to ask yourself some *practical*, and some *philosophical* questions about your work. These questions are sometimes called '5Ws and an H':

- **Who** am I writing for? Who are the main role-players in the argument? Who is affected by the issue?
- **What** am I trying to do/prove/achieve? What kind of approach or format do I need to take? What is the core problem?

- **When** do I need to submit my assignment? When (i.e. past, present or future) is the issue likely to be most apparent or critical?
- **Where** do I need to submit my work? Where does the problem originate from? Where is the problem at its most extreme?
- **Why** does my lecturer want me to answer this question? Why does the problem exist? Why do I think that my proposed solution might work to solve it?
- **How** formal or informal does my style of writing need to be? How objective or subjective does my writing need to be? How is the problem significant? How do I think the problem can be resolved?

Establishing answers to these questions will provide you with a more global picture of your overall purpose, as well as insight into your audience. In this regard, you will need to think about that first 'W' question more closely, and figure out exactly who you are writing for.

4.3 Analysing your audience

Knowing who your audience is helps you to figure out which content you need to include. It also allows you to select a suitable level of language and formality for your task.

For example, let us assume that Roberto's task required him to apply sources of law to a set of facts to answer a jurisprudential question. His audience, in this case, is his Jurisprudence 110 lecturer. It is reasonable to expect her to have a good command of legal language (often referred to as *legalese*) and to understand the legal content that he will be dealing with. So Roberto should be able to select law-specific sources, and use a fair amount of legalese to explain his argument. If, however, Roberto's task were to explain the same concepts to someone without a legal background, he would need to restrict himself to simpler language and concepts, or perhaps explain each concept used. But, because his audience is his lecturer, it is suitable for him to use a more formal and technical style.

> **Jurisprudence** refers to the study, philosophy, and theory of law. For example, 'Sisonke is thoroughly enjoying the theories of law that she is studying in Jurisprudence 110.'

In short, you can expect your reader to be able to read and attempt to understand your message, but you also need to play your part by ensuring that you address the two main needs of your audience:

- Firstly, the reader's **academic expectations**, which will require that you use a clear, high-quality argument, and that you convey relevant content from credible sources; and
- Secondly, the reader's **emotional expectations**. Although this element may sound strange for the purposes of a formal assignment, it is still a vital component of considering your audience. You need to select a suitable style that neither patronises, nor speaks over your reader's head by using impossibly technical or academic language. If you annoy your readers by coming across as superior, or by making them feel stupid, it does not matter how clever your argument is – your assignment runs a real risk of not getting the merit it deserves.

> In this context, to **patronise** means to speak condescendingly, or to speak down to someone in a way that makes them feel inferior or uneducated.

Whether you are preparing for a written or an oral presentation, you can ask yourself the following questions to help you to prepare a suitable approach:

- How many individuals will be reading/watching this?
- How familiar are my readers/viewers with the topic? (i.e. what educational background does my audience have?)
- What demographic segments do my audience represent, and what, if any, influence will this have on my approach?
- Does my audience think my topic is important?
- Does my audience have a particular interest in the topic that I would need to consider in my writing?
- Does my audience understand the problem?
- Do I know which sources my audience finds credible?
- Do I know which supporting points will have the most impact on my audience?
- What areas of my argument is my audience likely to disagree with?
- What kind of tone and language would my audience be most comfortable with?

 Visit the Learning Zone for an extended writing activity: good practice for analysing your audience and knowing your purpose (Student exercise LZ 4.1).

4.4 Facts, theories, opinions and objectivity

As you noticed in Figure 4.1, you will frequently need to adopt a more formal, objective approach in an academic environment, whether you need to persuade someone or convey information or opinions. This section will help to clarify what it means to be objective so that you can develop logical, appropriately supported arguments.

Before we progress to an explanation of objectivity, it is important to be clear on a few academic terms that are related to the concept of objectivity. We will also outline the roles that fact, theory, opinion, and subjectivity play in argumentation.

4.4.1 Fact vs theory

In scientific reasoning, something may be called a 'fact' only once it has been objectively proven through experimentation, careful observation or measurement. The problem, however, is that facts are frequently confused with theories. Let us take a look at a particular example related to global warming to illustrate these concepts:

- **Fact:** The global surface temperature of the Earth has steadily increased since 1880, and rapidly increased since 2005 (NASA, 2019)
- **Theory:** Humans have played a significant role in the observed warming at the surface of the Earth through their influence on greenhouse gas emissions (Cook et al., 2016).

If we look at our claimed fact, NASA and a number of other organisations have carefully observed and recorded the Earth's surface temperature, using reliable measurement instruments over an extended period of time. As such, various research centres can **corroborate** each other's' findings and agree that the Earth's surface temperature has, in fact, increased steadily since 1880, and more rapidly since 2005. Not too many arguments there.

What *is* sometimes in dispute, however, is the role that humans play in this observed temperature increase. *This* is where theory comes in.

Your average Citizen Jo might see this as 'just a theory' – meaning a basic idea or thought with little evidence to back it up. But in academic use, a **theory** is actually a rigorous, coherent, and systematic set of ideas that is testable, **falsifiable** and which is repeatedly tested to establish support for it (Bradford, 2017).

An academic theory thus tries to predict or explain a phenomenon, and will usually be discarded or changed if observation or practice finds that it is incorrect or incomplete. As such, the more we experiment using a theory and *fail to reject or disprove it*, the stronger it generally becomes.

Citizen Jo might say, 'Bah, that's just a theory' when they are trying to ignore a claim they do not like, but the reality is that theory – in the academic sense, at least – should be very well-supported with evidence.

The theory we cited in this section, for example, is the anthropogenic climate change theory, which argues that humans have driven the increases in the Earth's temperature. This theory is supported by similar findings, scientific experimentation and climate modelling by a majority of climate experts and organisations (NASA, 2019; Cook et al., 2005), and so the theory may well offer valuable insights into the causes of the increase in Earth's surface temperature. We should therefore be inclined to accept the theory. Because it makes particular predictions, we can use it to try to make the necessary, and now urgent, changes for the future.

Despite this scientific consensus, the human-based climate change theory is still rejected by some (including, sadly, some world leaders like US President Trump). Those who do not wish to acknowledge the extent of humanity's impact on Earth's climate may argue, for example, that global warming is just part of a natural cycle. This argument actually has an element of truth to it since climate change consensus (NASA, 2019) *does* include natural warming cycles as a factor in its modelling – the overwhelming consensus, however, is that humans are currently driving climate change (Cook et al., 2016). Another approach to rejecting climate change, which may be more disheartening, is for certain sceptics (and here again US President Trump is a good example) to confuse climate with wet or extremely cold weather in order to 'prove' that there is no such thing as global warming at all! In this case, there are no facts, theories, or even bits of theories to support this claim, so we should reject it outright.

In short, a fact is something that we can observe and prove objectively. A theory is something that cannot be proven, but which can be

To **corroborate** means to confirm or support what someone else has claimed. For example, 'South Africa's Meteorological Society corroborates NASA's figures indicating that global surface temperatures has increased over the last century.'

Falsifiability is the idea that for a theory or hypothesis to have weight, it must be disprovable (or falsifiable) before it can become accepted as a hypothesis or theory. Philosopher Karl Popper argued that no theory is completely correct, but if it is not found to be false, then it can be accepted as truth. For example, 'Anton concluded that nobody had been able to disprove the theory of evolution, despite it having many falsifiable elements, and so, as far as theories go, it is likely to be true.'

Visit the Skeptical Science website for more climate change myth vs science arguments at https://www.skepticalscience.com/argument.php if you are interested in the topic.

A great example of how to make a difference through acting on reliable evidence is Greta Thunberg, a Swedish teenager with Asperger's Syndrome who is on a mission to get world leaders to act on climate change. Read more in a *Time* magazine article by Suyin Haynes: <https://time.com/collection-post/5584902/greta-thunberg-next-generation-leaders/>, or follow her on Twitter (@GretaThunberg; Greta#Thunberg) or Facebook (@gretathunbergsweden) disproven or corroborated, and further supported with evidence. The likelihood that a theory will change, develop, and evolve, therefore, is high, and, in academia at least, a theory is never 'just a theory' in the Citizen Jo sense of the word.

4.4.2 Opinion

The difference between fact and opinion is one of the key differences that you will need to understand when you are engaged in writing and critical reading activities. As we said earlier in Section 4.4.1, a fact is something that we can prove. In the case of an **opinion**, however, we cannot prove that it is true or false, and the statement simply expresses a belief, feeling, or value on a particular topic.

While some might claim that we are entitled to our opinions, author Harlan Ellison (1996) bluntly disagrees, arguing that:

'We are not entitled to our opinions. We are entitled to our informed opinions. Without research, without background, without understanding, it's nothing. It's just bibble-babble.'

Ellison's own opinion on opinion may be a good way to keep your academic writing in check. If your opinion is informed (and supported), your arguments have a much better chance of being accepted.

Let's take a look at how an opinion differs from a fact by extending our earlier example about climate change:

> 'Anthropogenic climate change theory might have
> scientific support, but I think that it's all a conspiracy
> by the EU to achieve global domination.'

This statement includes both a fact and an opinion. Can you identify which is the fact, and which is the opinion?

It should not be too difficult to find the factual statement – you simply examine the statement to see if it can be proved. Let's take a closer look:

> 'Anthropogenic climate change theory might have
> scientific support, but I believe that it's all a conspiracy
> by the EU to achieve global domination.'

For the first part of the statement, can we prove that anthropogenic climate change theory has scientific support? Absolutely – even a quick Google-search of *credible* sources would confirm that this is the case (don't worry if you're not too sure about fact-checking yet – we will work on your 'Google-Fu' in Chapter 6).

The next part of the statement, however, needs a little more work. We *cannot* prove that anthropogenic climate change theory is part of a conspiracy by the European Union (EU) to achieve global domination – although we may indeed find similar claims of this nature on some *less credible* and/or conspiracy websites. Since we cannot prove the statement from our personal experience and observation, we would have to conclude that this is not a factual statement.

Apart from the quick fact-check approach to analysing statements, you can also ask yourself a few questions to determine whether or not the statement is opinion:

- Does this describe a thought or feeling? (If it does, it's an opinion);
- Will the statement be true all of the time? (If it will not be, it cannot be a fact);
- Are there any judging words or words that suggest an opinion, such as 'feel', 'believe', 'always', 'never', 'none', 'worst', 'unfortu- nately', etc.? (If you see one of these words, the statement is likely an opinion).

As you can see from these examples, opinions often provide clues that point to the person expressing them, such as 'I think', 'I believe', so they should be relatively easy to spot. Unfortunately (see what I just did?), they are also often 'smuggled' into academic writing where they have no real business.

For example, if an author were to comment on the '… <u>unacceptable</u> refusal of governments to accept the anthropogenic climate change argument', they would be injecting their own feelings about a topic into the argument. The opinion may well be valid, but by including it here, the author undermines any scientifically supported statements made alongside the opinion.

Therefore, if opinions are included at all, it is wise to ensure that they are well-supported by solid reasoning and/or evidence.

 Visit the Learning Zone to complete an activity that will help you to identify the differences between facts, opinions and theories (see Student exercise LZ 4.2).

4.4.3 Objectivity and subjectivity

As we have already mentioned, much of the academic writing that you will be required to submit during your studies will require an objective approach. But what does it mean to be objective?

Although there are varying degrees of objectivity, if you take an **objective** view in your writing, it means that you try to:

- Avoid emotive language and be as neutral, or unbiased, as possible;
- Present facts and evidence, which you allow to speak for themselves without injecting your opinion or personal values into the narrative – i.e. the truth is independent of your existence or commentary;
- Use verifiable statements from credible sources;

- Be as specific and explicit as possible (e.g. 72% instead of 'most of the …');
- Use the third-person (e.g. he/she/the researcher/Einstein etc.) to convey ideas;
- Present balanced argumentation;
- Avoid intensifiers like 'very', 'awfully', or 'really', which exaggerate writing;
- Avoid language that excludes any group of people;
- Use modality to show caution about your views (e.g. '… evidence suggests that this *may* be a cause …');
- Observe the scientific method as far as possible.

Subjective writing, on the other hand:

- Presents your opinion, viewpoint, character or values instead of facts – i.e. they reflect your idea of the 'truth', and not an externally verifiable one;
- Often provides an incomplete, general or narrow picture on a topic;
- Uses the first person ('I' or 'we') or second person ('you') to convey ideas;
- Often makes use of intensifiers like 'very', 'awfully' etc.;
- May make use of absolute statements (e.g. 'The study shows all humans activities cause …');
- Does not use a scientific method;
- May include statements that are partially, or even entirely true, but that are not necessarily so, and more often include personal viewpoints.

 Visit the Learning Zone to practise your skills in identifying objective and subjective writing samples, using Student exercise LZ 4.3.

Depending on which faculty you are studying in, and what kind of assignment you have been given, you may be tasked with using either subjective or objective writing. As we've already mentioned, most of your academic writing will require a more formal, objective style. Some reflection activities and social science studies do, however, require that you provide your own insights and opinions on different topics, so do not discount the subjective approach completely.

4.5 Writing styles and formality

Tone refers to the attitude that a writer displays to their audience or topic, and is conveyed through word choice, what they choose to include, and through their manner of expression. For example, 'Andile's playful tone came through in his writing.'

For your current academic purposes, writing style and register refer to the way in which you use words to convey meaning, or 'a way of writing or speaking', as Fielding and Du Plooy-Cilliers (2014:162) put it. Your language complexity, sentence structure, objectivity, vocabulary, formality, and choice of tone will all contribute to a particular style, so you may, for example, make use of a social style, or a more formal style, depending on what you are trying to achieve.

The style that you select will be strongly influenced by the level of formality required. Generally, the more objective your writing needs to be, the more formal it will be, and vice versa. Examine Table 4.1 below, which outlines some of the main styles or registers that you might use in your studies.

Table 4.1. Writing styles or registers

Style / Register	Appropriate for	Characteristics
Professional	*Writing for a particular profession (e.g. law or medicine)*	• Extensive use of jargon specific to the group being engaged; • Not necessarily formal but relies on shared expertise and understanding of the professional language.
Formal	*Writing complex, academic pieces such as dissertations, but may also be used for many kinds of published or business documents, depending on the degree of formality chosen*	• Does not try to develop a relationship between the writer and reader; • Uses formal vocabulary that relies on broadly accepted definitions – contractions and slang are never used; • Demands use of standard English grammatical structures; • Uses third person (impersonal he/she) at its most formal, but may be relaxed to first person (I/we) depending on circumstances and audience; • Employs wordy, complex sentences and paragraphs, and often uses the passive voice, in its most formal version, but if a slightly lower level of formality is allowed, concise sentences and paragraphs in the active voice may be used; • Objective – uses concrete, referential words and facts, and avoids emotive language; • Uses substantiated argument and fact when making statements.
Casual	*Used for informal communication in social situations*	• Depends somewhat on shared experiences, interests and/or even personalities; • Does not require standard English grammatical structure; • Often makes use of unique code or language (slang) that only has meaning within the particular group; • Usually takes a friendly tone but is suitable for expressing a range of emotions; • Wholly unsuitable for academic work – leave this register for socialising!
Intimate	*Used within your inner circle of closest friends and confidants*	• Highly dependent on shared experiences; • Does not use standard English grammatical or syntactic structures; • Assumes an understanding of language and words that may be unique to the members; • Like the casual style, do not use this for academic work.

You may have noticed that we have not discussed some of the main elements of style that are referred to in Table 4.1 above. Let's take a quick look at these elements now:

4.5.1 Active and passive voice

The English language makes use of two main 'voices', namely active and passive. The **active voice** directly engages with your audience by making the 'actor', or subject of the sentence, perform the action.

For example, in the sentence *Einstein developed the general theory of relativity* the subject of the sentence (Einstein) has actively performed the verb (developed). Therefore, the sentence is said to be written in the active voice.

In the **passive voice**, the object of the sentence becomes the focus, and we remove the 'actor' to some degree. If we reword the above example so that it is in the passive, we would say: *The general theory of relativity was developed by Einstein.* As you can see, the sentence still means the same thing, but it becomes longer, and does not engage directly with the audience.

While active voice is recommended for speeches and other forms of writing, you use the passive voice in academic assignments if you need to:

- use an objective, impersonal style;
- remain neutral or tactful by avoiding 'blaming' someone;
- emphasise the object of your sentence.

It is important to note that using the passive voice makes your writing less readable, and adds significantly to the complexity of your writing. If you do decide to use it, ensure that you proofread your work for consistency (it is very confusing to the reader if you switch between the active and the passive voice within a sentence or a paragraph).

4.5.2 Referential language

Table 4.1 also states that the formal style makes use of **referential** language, and avoids emotive language. In simpler terms, this means that objective academic writing tries to make use of factual (referential) words more than words that evoke emotions in a reader.

Remember that, in addition to using referential language, you also need to use suitable, credible sources in your academic writing.

4.5.3 Jargon

Jargon refers to a particular vocabulary that a group of specialists or professionals use to communicate with each other. It is exclusive, by nature, and someone without similar experience, background or skills may not understand the language of the particular group.

For example, if a group of surgeons is discussing an operation that they are going to perform, it is unlikely that most of us would understand their conversation very well. Similarly, economists, lawyers, sportspeople, and academics have jargon that they use when communicating with each other.

Although there is nothing wrong with using jargon when you are sure that your audience understands you, it is vital that you consider your audience first. For example, if there is a chance that your readers will not understand your terminology, then you either need to explain it (if you *must* use the terminology), or try to avoid it altogether.

4.6 Improving readability

Because you are likely to use a more formal register for most of your academic writing, it is useful to establish some of the main elements that can help to improve the readability of your text. **Readability** is the ease with which someone is able to read and interpret your work. For example, writing that flows naturally and that uses simple sentences and signposts different ideas is easier to read than writing that uses complex sentences, technical words, and abstract concepts. Depending on the purpose of your writing and the audience that you are addressing, different readability levels are suitable.

We can measure readability manually, by using a formula and counting words, sentences and polysyllabic words, but word-processing software can quite easily do this for us. Microsoft Word, for example, can provide you with readability statistics on two scales or indexes:

- the Flesch Index measures readability on a scale of 0 to 100, with 0 being the most difficult; and
- the Flesch-Kincaid Index measures readability according to the educational grade for which it would be suitable. The lower the grade, the easier the piece is to read.

 Head to the Learning Zone to see how these readability measures work in practice! (See Student exercise LZ 4.4.)

For academic essays, papers, and journal articles, you should try to keep the readability score of your writing between 30 and 50 points on the Flesch reading index. This level of writing is considered quite difficult to read, but given that your audience is most likely a lecturer who is familiar with your subject matter, it is likely to be a suitable level for your task. Chapters 7, 8 and 11 will help you to write in a style that yields this level of readability.

For more formal scientific papers or academic theses, you may consider aiming for a readability score of between 0 and 30 on the Flesch reading index. There are academic situations in which this would be suitable, although this is not generally advised, especially at the undergraduate level.

Some of the elements that affect the readability of your work include:

- **Layout**
  - > Follow the format prescribed by your faculty or as instructed for the task;
  - > Use a coherent, organised structure;

In the context of writing, a **signpost** refers to particular words or phrases that either indicate the direction of a thought or argument, or link and show connections between different ideas in writing. For example, 'This essay critically examines ...' is a good example of a major signpost as it clearly signals a key aspect of the work. Similarly, if you used a string of 'firstly ... secondly ... finally' etc., you would be using linking words to signpost your argument in a logical manner. Good signposting helps a reader navigate or find their way through your argument. You will learn more about signposts in Chapters 7, 8 and 11.

Polysyllabic words are words with more than two syllables. For example, if you sound out the word 'readability', it consists of five syllables. It is therefore a polysyllabic word.

Note!

Whatever level you pitch your writing at, do not fall into the trap of trying to 'sound clever' by using overly complex language that you do not understand. While it is important for you to enhance your vocabulary, language and writing skills during your studies, it is equally important that you do not use language without fully appreciating its meaning.

> Use headings and subheadings to signpost your thoughts (where necessary, and only if the format you were given allows for this);
> Ensure that your paragraphs include one topic sentence and a sufficient number of supporting sentences;
> If necessary, use a decimal numbering system or indented sub-sections;
> Do not use 'gimmicky' fonts like Comic Sans or French Script, even if you think they make your writing look pretty – stick to fonts like Tahoma, Times Roman or Arial, in a size that is large enough to read;
> Include a high percentage of white space on your page by using a double-spaced paragraph setting, and margins of at least 2 cm on all sides (if the instructions for your task allow this).

- **Word choice**
  > Make sure that your vocabulary is at a level that your audience understands;
  > Only use jargon if you are sure that your audience will understand it – where necessary, provide definitions if you do use it.
- **Style**
  > For most academic pieces, you will use a formal style – this will, however, reduce the readability of your writing, so remember to bear your purpose and audience in mind when you select the different elements of style to decide on your approach.

The number of these elements may appear a little overwhelming at first, but you will start to notice how they 'click' into place quite easily as you practise using them in your writing, and once you get the hang of things. These elements are intimately related to each other, so don't worry if you have not got them all working just yet. Practise, practise, and practise a little more, and you will start to use a consistent, formal academic style that effectively conveys your research.

 For lecturers, there is a class group activity available on the Learning Zone, bringing together all that students have explored in this chapter.

4.7 Conclusion

This chapter outlined the various purposes of written and oral tasks, and provided a simple method for analysing your purpose and audience. Using a practical example, it examined the differences between some of the most important academic concepts that influence the register and style of your writing, and offered guidance for improving the readability of written work for academic purposes.

References

Bradford, A. 2017. 'What is a scientific theory?'. *LiveScience*, 29 July 2017. Retrieved 30/11/2019, available at <https://www.livescience.com/21491-what-is-a-scientific-theory-definition-of-theory.html>

Cook, J., Oreskes, N., Doran, P. T., Anderegg, W. R., Verheggen, B., Maibach, E. W., Carlton, J. S., Lewandowsky, S., Skuce, A. G. and Green, S. A. 2016. 'Consensus on consensus: a synthesis of consensus estimates on human-caused global warming'. *Environmental Research Letters*, 11(4). Retrieved 17/00/2019, available at <https://iopscience.iop.org/article/10.1088/1748-9326/11/4/048002>

Ellison, H. 1996. 'Harlan Ellison's Watching'. *Sci-Fi Buzz Commentary*, n.d. Retrieved 30/11/2019, available at <http://harlanellison.com/buzz/bws006.htm>

Fielding, M. and Du Plooy-Cilliers, F. 2014. *Effective business communication in organisations: Preparing messages that communicate*. 4th ed. Cape Town: Juta.

Haynes, S. 2019. '"Now I am speaking to the whole world." How teen climate activist Greta Thunberg got everyone to listen'. *Times Magazine*, 16 May 2019. Accessed 18/08/2019, available at <https://time.com/collection-post/5584902/greta-thunberg-next-generation-leaders/>

Hemingway, E. 1932. *Death in the afternoon*. New York, NY: Charles Scribner's Sons.

NASA. 2019. 'Facts: Scientific consensus: Earth's climate is warming'. *NASA*. Retrieved 29/08/2019, available at <https://climate.nasa.gov/scientific-consensus/>

Skeptical Science. 2019. 'Global warming and climate change myths'. *Skeptical Science*. Retrieved 20/08/2019, available at <https://skepticalscience.com/argument.php>

CHAPTER 5

Digital citizenship

'Is this technology serving me, or am I serving it?'
– Nir Eyal (2019)

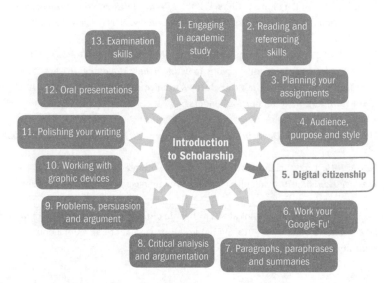

1. Engaging in academic study

2. Reading and referencing skills

3. Planning your assignments

4. Audience, purpose and style

5. Digital citizenship

6. Work your 'Google-Fu'

7. Paragraphs, paraphrases and summaries

8. Critical analysis and argumentation

9. Problems, persuasion and argument

10. Working with graphic devices

11. Polishing your writing

12. Oral presentations

13. Examination skills

Introduction to Scholarship

OBJECTIVES

At the end of this chapter, you should be able to:

- outline the importance of digital literacy and citizenship;
- identify the elements of digital citizenship;
- discuss contemporary concerns about ethics in cyberspace;
- identify practical ways in which to manage ethical concerns in cyberspace;
- use online tools to fact-check news or social media items;
- use advice and tools to demonstrate sound digital citizenship.

Vhudihawe is a first-year BSc student, and she is worried. At first, she and her friends thought the single rose on her desk every Biology lecture, along with an occasional, rather mushy, anonymous love letter was cute: a minor mystery that demanded little more than a laugh over coffee. Three weeks later, she is no longer amused. After a night out with friends, she has found a rose and a long letter – a letter describing all of the ways that she is the embodiment of perfection: her beauty; her brains; her empathy for others; her laugh; her dance moves. At. Her. Front. Door. She has no idea who is sending these messages or flowers, but they seem to know everything about her – including, most frighteningly, where she lives. She is scared.

Her friends convince her to report the issue to campus security. After speaking to them, she feels even worse. After asking a few questions and doing a quick check on a computer, the officer shows Vhudihawe just how much of her life is public. Between Facebook, TikTok, and Instagram, Vhudihawe realises that her life is an open book, and that her 'secret admirer' does not need to be following her physically to know where she lives, learns, hangs out, or plays. The officer expresses sympathy and concern, but tells her that since no crime has been committed and the mystery person remains a mystery, there is little that campus security can do. For now, she needs to be vigilant, keep her doors locked, avoid walking alone around campus and – if she wants to avoid this happening again – change the privacy settings on her social media apps immediately. Despite acknowledging that she has not been careful enough on social media and needs to fix this, Vhudihawe is still angry that all the 'solutions' centre on changing *her* behaviour, and that nobody will do anything about the behaviour of the person stalking her. It should *not* be like this!

5.1 Introduction

Vhudihawe's situation, while *possibly* emotionally distressing but physically harmless, could easily become a dangerous one. Unfortunately, this IRL is more realistic than hypothetical, as reports on gender violence, and social movements like *#AmINext*, can attest. It is *also* true that it is often law-abiding citizens, rather than those who intrude upon them, who adapt their behaviour to be safe and private in what is a public, messy, and potentially dangerous world.

Maintaining privacy, and the security that goes with it, is therefore critical. It is also just one of the many digital issues that students need to be aware of today. There is growing agreement that we are moving out of the information age and into a new and even more complex era – one of innovation, experience, and, perhaps more scarily, reckoning (Birkinshaw, 2018; Wadhera, 2016; Doria, 2014). It is no longer enough that we use digital platforms to create and share information – we now need to *do* something new and creative with it, at the same time as acting in a safe, ethical, and environmentally sustainable way.

If we fail to opt in to this shift in thinking, we risk being left behind: not just socially ('What? You don't have an Imgur account?'), financially (we have some rather ... um ... entrepreneurial fraudsters in this country,

Take the Internet and online communications quiz on the Learning Zone (LZ 5.1) to get your mind moving in the right direction for this chapter.

Retrobate is a colloquial term in online communities for someone who refuses to adopt new technologies and skills, and insists on doing things 'old school' – more formal language might refer to this person as a Luddite.

so be careful!), and educationally (lifelong learning is going to be considerably harder if you do not know an LMS from a bookmark), but also in terms of employability. Unfortunately, 'Now hiring retrobates' signs are few and far between in southern Africa. Not to put too fine a point on it: we urgently need to embrace being digital citizens.

This chapter will offer insights into what it means to be a digital citizen and will examine some of the core cyberethical concerns of today. Even if you are a digital native – someone who has been brought up in an age of technology and computers – you may never have been taught the skills and thinking to manage your online life safely and constructively. If this sounds unimportant, consider Vhudihawe's experience – being stalked is just *one* of many possible implications of living when social media is pervasive. The list of possible negative consequences is long indeed. That said, there are also many benefits to opting in, so in offering you some practical steps and advice towards responsible digital citizenship, we can try to ensure that technology is serving you, rather than the other way around!

If something is **pervasive**, it is spread everywhere – in this case, our reliance on the World Wide Web and technology is so widespread that it would be almost unimaginable to live without them.

Point of interest

The UK's Royal Society for Public Health (2017) found that social media use by teens produces effects ranging from improvements in self-expression and community building on the positive side, to depression, bullying, and sleep deprivation on the negative side. For tips from local and international celebs on how to use social media in a healthy way, visit SADAG (The South African Depression and Anxiety Group) at http://www.sadag.org/ and search "celeb tips social media".

5.2 Elements of digital citizenship

You have probably come across the words 'digital citizenship' and 'digital literacy'. The terms are closely related, but what exactly do they *mean*, and why are they important?

For the purposes of this book, we'll accept a definition of 'digital literacy', although the concept seems to evolve almost as quickly as technology itself. Beetham and Sharpe (2011:1, in Spante et al., 2018) argue that to be **digitally literate**, you need to be an 'agile adopter of a range of technologies for personal, academic and professional use'. This means that you need to be able to get used to and use different hardware and software, online platforms, and apps in order to find information, and decide what is credible and what is not. You *also* need to be able to use that information responsibly and productively.

On the other hand, **digital citizenship** is a much broader concept – it refers to how confident a person is in their ability to thrive in a digital world. Ribble (2017) describes digital citizenship as the ongoing development of a set of skills and norms that allow users to feel empowered by technology. The recognition that learning is ongoing is important. So is the recognition that learning relies on understanding norms and ethics as well as skills. This means that we should remember to use technology for the greater good, and not to the detriment of others.

Ribble (2017) notes that being a good digital citizen requires a person to be competent in nine elements:

1. Digital access relates to our dedication to expanding access to technology and online tools – for everyone, regardless of their financial or social circumstances;
2. Digital commerce aims to ensure that we have online financial and security savvy;

3. Digital communication focuses on how well we can express and share our ideas across different platforms;

4. Digital etiquette (sometimes called 'netiquette') obviously expects us to mind our 'P's and Q's', and be polite in the digital space;

5. Digital literacy includes a number of elements related to practical online skills, including the ability to distinguish between credible information and unreliable information, and to become as digitally fluent as possible;

6. Digital health and welfare, relates to the ability to use digital and online tools in a way that positively – rather than negatively – impacts our psychological and physical health (um … like not spending too much time on our phones!);

7. Digital law includes understanding and respecting legal responsibilities, policies and actions online – for ourselves and for content creators – so that we know our rights and stay within the law;

8. Digital rights and responsibilities are our rights, freedoms and responsibilities, and how we can use these constructively in a way that protects and helps both ourselves and others;

9. Digital security and privacy refers to knowing how to avoid those nasty little viruses, worms, and bots that get passed around online.

Feeling overwhelmed at how many elements there are to navigate? Don't worry – if you try hard, you can probably think of one positive example for each of the nine elements already – so what follows is probably not all new to you. Sections 5.3 and 5.4 will take you through issues and tricks to make sure you feel comfortable stamping Digital Citizen on your passport!

5.3 Cyberethics

The Internet has arguably become the most widely used information tool in society – as a result, *ethical* issues are bound to surface. Issues around privacy, confidentiality, piracy, plagiarism, big data, bias, fraud, intellectual property, cyberbullying, and even fake news are common concerns relating to the World Wide Web (Zimmer, 2018; Buchanan and Zimmer, 2018). Many of these relate specifically to *conducting* scientific research using the Internet, and so will not feature in this particular discussion, but it is important to recognise and grapple with some of the relevant ethical issues that you may face in your undergraduate years.

This section briefly examines seven areas of ethical concern that you may encounter in your studies, including privacy, search engine bias, freedom of expression, the digital divide, intellectual property, fake news, and cyberbullying.

5.3.1 User privacy

Privacy slips on social media sites like Facebook, Imgur, and TikTok can be relatively minor (socially embarrassing), but if you've published

> **Point of interest**
>
> For a fascinating – if slightly frightening – look into the digital skills gap and future of jobs, watch Growth Tribe's (2017) video at <https://www.youtube.com/watch?v=Y9FOyoS3Fag>

Ethics is a branch of moral philosophy that investigates questions of right and wrong, or 'good' and 'bad'. Note that the word 'ethical' means something different to 'legal'; the word 'unethical' also has a different meaning to illegal. For example: Katlego's boss has asked her to print 20 copies of a PDF she found on the Web so that she could use it to present a training course and claim it as her own. Katlego knows that this is wrong, but she also does want to get into trouble with her boss, so she has an ethical dilemma on her hands.

something incriminating, they can also cause you trouble when you are looking for a job, or when you are employed. For example, we doubt a potential employer will appreciate the time you (literally) painted the town red with spray paint while posting selfies of your obviously drunken face. The consequences do not end there. Privacy issues can also result in situations that are potentially more dangerous, such as the one Vhudihawe finds herself in, or in other serious crimes like identity theft or fraud.

Potential invasion of privacy is one of the greatest concerns that critical users of the Web have. This can happen in a number of ways, but careless privacy settings, big data breaches, search engine 'bots', and other spyware are by far the most common.

Firstly, failing to set strict limits on who can see your profile and posts on social media apps is a simple but common way we forgo our privacy. When our information is then accessed by someone other than ourselves, the purpose may, or may not, be to cause deliberate harm or gather sensitive information about us.

A more deliberate invasion of privacy occurs in the case of a data breach. This is a security incident in which a hacker harvests personal and supposedly confidential information (Ramsey, 2019). Cyber criminals seem to have recently increased their attacks on South African (and global) companies' data, leading to personal details (like ID numbers, addresses, and even salary slips) on servers of organisations as varied as Facebook, Ster-Kinekor, Liberty, and Jigsaw Holdings being 'harvested' (Niselow, 2018). There is no guarantee – thank goodness – that anything fraudulent will be done with your information even if it is in the hands of hackers, but the threat remains. As customers of these companies, there is almost nothing we can do about this kind of intrusion, but companies must work very hard to restore customers' faith in them. They otherwise risk losing customers.

A third way that your privacy might be violated is through 'bots' and spyware. These are also used to dig into your – apparently very desirable – privacy. For example, search engines can collect your personal information, details of the device you are using, your phone number, login information, search history, location, details of phone calls, and information about which videos and sites you click on (Google, 2019). Just through using sites like Google, you agree to – and even encourage – this invasion. Before you decide to swear off 'Big Brother' technology forever though, note that giving search engines access to all of this information *does* help to make your life easier. For example, Google (2019) explains that this information helps it to:

George Orwell used the term **Big Brother** in his book *Nineteen Eighty-Four* (1949) to refer to the ever-watchful eye of the government.

- improve and develop new services;
- provide personalised services, including content and advertising;
- ensure that things are working as they should be;
- communicate with you about security threats or changes to services;
- reduce spam and harmful malware, and therefore protect you; and
- combine information from one service with information from other services.

By allowing search engines to use personal information, you help them to cater to your personal needs – this definitely speeds up your searches and helps to limit the number of completely irrelevant adverts and suggestions. Just be mindful, check your privacy settings, and know what you are getting into if you continue to use 'Professor' Google as your BSEF (best search engine forever). If you're still troubled by Google's approach, we suggest some alternatives for you in Section 5.4.1.

5.3.2 Search-engine bias

Search-engine bias is an ethical issue that has a direct impact on your use of the Web. This bias refers to the fact that search engines are not neutral when looking for matches to your search phrase; instead, they favour some values over others. Kassner (2013) explains that the results of an online search are not objectively generated. For example, if two different people on two different devices type in exactly the same query, they are likely to get similar, but visibly different results.

This is because search engines are editing, and therefore deciding, which information they think each user will find more valuable. So, if one of these individuals frequently visits online tabloids and (ahem …) less academic content, that person may be directed to a very different, less academically suitable, set of results than the other person, who does more searches for academic content. As a result, each person's perception of the available content online may differ.

This may sound like a small problem, but it could mean the difference between finding excellent sources for your assignment, or only finding mediocre (or even terrible) ones!

It is important to note that these differences in results are due to search engines tailoring content for their users, so don't think that they are deliberately trying to sabotage your academic career. Experts like Kassner (2013) do not argue against using search engines, but they do argue for being aware of the built-in bias of the search engine system. If you are aware of this bias, you can adapt your search habits accordingly, as we will explain in Chapter 6.

It may be important to consider here that companies like Google (and Facebook and Amazon) are becoming ever-larger monopolies. This means that Google is managing more and more of our data, and we've given it almost unlimited permission to track our online habits. It is now in a superb position to subtly (or not so subtly) influence our thinking and actions through the search results it delivers. Think about this, and again, if you'd like a few alternatives, visit Section 5.4.1.

5.3.3 Freedom of expression – within reason

Section 16 of the South African Bill of Rights guarantees its citizens the right to '… freedom of expression, which includes freedom of the press and other media; freedom to receive or impart information or ideas; freedom

If you are **biased** against someone, it means that you treat them differently from others for an unfair or incorrect reason. For example, 'That umpire was clearly biased against the visiting team.'

To be **objective** means to consider all facts or elements without allowing personal feelings or opinions to interfere. For example, 'The judge managed to make an objective ruling despite the public emotion around the case.'

of artistic creativity; and academic freedom and freedom of scientific research' (Government of South Africa, 1996).

This means that South Africans are afforded a large degree of space in which to think and express our opinions on any platforms available to us. Importantly for students, it also means that we can access a wide variety of opinion and be influenced by many different ideas when we conduct searches using the Web (or any other means). Many countries in the world do not provide for freedom of speech or access to information in the same way that South Africa does, so we are in a particularly fortunate academic situation.

However, remember that freedom of speech is limited in certain ways. For example, it prohibits citizens to exercise freedom of speech if that speech:

- extends to propaganda for war;
- incites imminent violence;
- advocates hatred based on race, ethnicity, gender, or religion, and constitutes incitement to cause harm.

From a legal point of view, using platforms like Facebook or Imgur to express personal views and behaviours should reflect a *constructive* view of the world that does not violate any of these limits. Criticism of a constructive nature is perfectly acceptable – even desirable – in a democracy; indeed, one of the goals of your studies should be to develop effective critical-thinking skills. Sharing your opinion and backing it up with sound logic and evidence is a critical part of this process, and will also give you the 'higher ground', a position of ethical or strategic advantage, in any difference of opinion.

Yet personal attacks, or criticism without evidence or substance, are not constructive, and could find you in violation of the Constitution. For example, in early 2015, Speaker of Parliament Baleka Mbete used the word 'cockroach' to describe EFF leader Julius Malema. Since the word 'cockroach' was used in political propaganda during the 1994 Rwandan genocide in which almost a million people were killed, the Speaker ultimately acknowledged the harm that her words may have caused, and apologised unreservedly for her statement.

In this case, the Speaker may still have believed Julius Malema to be a 'cockroach', but, for professional and legal reasons at least, she apologised. Professionally, she is in a position of political power, and name-calling is not a professional way of behaving; legally, she might have run into trouble for what was clearly a violation of Section 16. By apologising, she avoided further actions, but the incident has negatively affected the way that some people view her.

In the same way, you may be explosively angry about something or someone, and wish to vent your rage on a platform you have access to. Try to consider the implications of what you say before you click 'post'! If you go ahead and rage, you may be causing someone personal harm, and also doing your own reputation considerable damage in the long run. South Africans like Penny Sparrow and Adam Catzavelos have learnt this the painful way, and have faced both legal and social consequences.

Using online and other platforms constructively is also important from a social point of view, and potentially even from an economic point of view. Because educational institutions are using online platforms more regularly, both for assessment and research, you may need to build an online 'presence' of your own as part of your studies.

This may include doing online research, developing an online portfolio, building a name for yourself on a particular forum, or even keeping a personal blog. Keep your communications within the ethical boundaries prescribed by your academic institution, and make sure that whatever you post can pass the '5-year' test. Ask yourself: if I saw this in 5 years' time, how embarrassed would I be? If you answer 'a little' or 'a lot', do not press 'post' – rather start again.

In summary, what you do and say in these online spaces is critical, as is the quality of information you quote: it could even be the difference between impressing a future employer and being excluded from a job opportunity. This may not be fair, but employers often do run checks on candidates' social media before interviewing. You should therefore either ensure that your personal social media has extremely high privacy settings, or that you avoid incriminating posts and images altogether. The better choice is to keep your posts constructive, and try to consciously build your image as a thoughtful and productive thinker. Your posts could even be the best CV you have ever had!

5.3.4 Digital divide

The **digital divide** is a term used to describe the information gap between people who have access to digital information technology, and those who have little, or no access to it (Internet World Stats, 2019). This divide may include a difference in actual access to computers, telephones, television, and the Internet, but it can also refer to the skills, or lack of skills, to access information technology even if it is available (Ibid.). So, if someone has access to an Internet-enabled computer in a local library, but does not know how to find information on it, one could say that they are on the wrong side of the digital divide.

This gap has the potential to widen the social, political and economic divides between those who have access, and those who do not (Rainie, 2013). Reducing the gap has therefore become a major focus of governments and non-governmental organisations around the world.

Advocates for reducing this divide argue that access to information technology can help to reduce economic inequality, improve social mobility, develop economies, and even play an important role in building healthy democracies (Internet World Stats, 2019). Since smartphones have become more widely available in southern Africa, the problem is more that people do not know *how* to access credible, useful information successfully, rather than not having access to it at all.

There has been a great deal of important debate on paid vs open access academic publications. As a student, you might conduct a search and find the perfect article to support your argument, only to find that you would have to pay US$40 to access it! To us, paying hundreds of rands for an unseen resource does not sound reasonable. As a result, some great resources effectively prevent you from using them, keeping you on the dark side of the divide. However, many journal articles you source will be free (open source, or licenced under Creative Commons), or available through your university's library database (having been paid for by your institution).

Van Noorden (2013) notes that those who favour charging for journal articles argue, perhaps reasonably, that editorial quality may be jeopardised in open source journals. For example, the more credible the journal, the more peer reviews it will conduct, and the more articles it will receive and reject, meaning that only the best articles will be presented to readers. Critics in support of paid articles argue that it costs money to run a journal, conduct peer reviews, and ensure that articles are properly edited, so it is unreasonable to remove reader payments (Ibid.).

Those in favour of open access journals to reduce the digital divide argue that charging fees for content further entrenches the divide, prejudicing those who do not have the funds to purchase this 'quality' content. There is also debate about the value that these 'quality' journals add (Van Noorden, 2013). As a result, many educational institutions, including prestigious universities such as Stanford, Harvard, MIT and Yale, have started offering MOOCs (Massive Open Online Courses). Students are also offered course material and multimedia through sites and apps like Udemy, Khan Academy, Codecademy, and Apple's free iTunesU Free Courses app (Friedman, 2019). Students are seldom awarded graduation certificates when learning through these apps or through MOOCs, but they are able to learn and be supported in their learning, and can engage in the marvellous practice of lifelong learning (Koller, 2012).

Debates around paid vs open access content will continue, and MOOCs may change to be more selective in future, or become ever more popular. For now, it is enough to realise that whichever side of the divide you are on, it is worthwhile to be aware of the divide itself, and then learn as much as you can from the resources you have available.

The Publishers' Association of South Africa has developed an easy-to-use Copyright Information Guide that answers a number of common questions about copyright. Visit the Learning Zone and click on the link to find it <http://www.publishsa.co.za/downloads/copyright_information_guide.pdf>.

5.3.5 Intellectual property

Intellectual property (or IP) is any product of the mind, such as an invention, book, song, or artwork; depending on their type, these creations can be protected by copyright, trademark, patent or registered design (Publishers' Association of South Africa, 2007). For example, an invention could be patented, but an academic journal article would be protected by copyright law.

For products that can be protected by copyright law, the creator holds the legal right to decide where or how that work appears. In many cases, a creator will license a publisher to make those decisions on their behalf. The good news is that copyright protection applies to everyone (you do not have to register your work to be protected), so if you create something original, you are also entitled to IP rights over that work or creation.

As a student, it is your responsibility to ensure that you respect the IP of others. Reading through and sticking to your educational institution's IP policy is one of the best ways you can do this. Note that some institutions may refer to this policy as their plagiarism policy, but protecting IP really goes beyond plagiarism, as IP can be used, and abused, in a number of ways.

Typically, online IP abuse includes misusing brands in order to commit fraud or sell fake products, or it may include downloading, uploading, altering, distributing or selling documents, video, images, or music (Schaefer, 2011).

In an academic context, the most notable ways in which copyright is infringed are through plagiarism in its different forms, through unlicensed or unreferenced use of images, and through the illegal (or pirated) printing of academic articles or textbooks. Copyright law does make exceptions for 'fair use' (Geach, 2014), but this term is subjective, very restricted, and may be subject to a number of complicated checks that often include applying for written permission from the copyright holder.

Fair use is usually understood to mean using copyrighted material in a very limited way, with acknowledgement of the copyright holder or licence holder, but without their permission. However, what this means in practice is not clear, and people can have different understandings. What seems 'fair' to a student might seem 'unfair' to a publisher or author.

The nature of IP can be difficult to pin down, as you can see. As a result, the laws around IP can be complex and even 'grey', especially since legal infringements are often determined on a case-by-case basis. On a more philosophical level, we would encourage you to continually reflect on how, when and why you use IP belonging to someone else while you are creating your own academic work.

From a practical perspective, the most important things are to comply with both ethical and legal requirements relating to the use of other people's IP, and to be able to present your arguments with a clear conscience. Some practical suggestions for how you can achieve this include:

- being extremely familiar with your institution's IP and/or plagiarism policy and practices;
- making sure you know who the author or creator of any IP is;
- where necessary, requesting permission to use the content, and paying any necessary fees to do so;
- making sure that you reference the content correctly;
- citing the author or creator wherever you have used his/her ideas etc.;
- running your academic 'creations' through your academic institution's plagiarism-checking software, if possible (Turnitin, SafeAssign, PlagTracker, DupliChecker or Viper are commonly used and available software programmes that do this).

Refer back to Chapter 2 of this book for step-by-step guidance on how to avoid infringing IP by citing your sources, and consult Chapter 7 for guidance on how to paraphrase suitably.

 For this section, we suggest that you go to the Learning Zone to complete a number of short, useful activities on IP to get a more concrete idea of the IP issues that you might encounter during your studies.

5.3.6 Fake news

By now, we are sure you have heard the term 'fake news'; you may well have also come to associate it with Donald Trump, the 45th president of the United States, who popularised the term (but did not coin it) (Wendling, 2018) by misusing it. The original meaning of fake news refers to news that is a hoax, or, at the very least, factually incorrect and which exists to mislead, damage, or make some political or financial advantage (Ibid.). However, Donald Trump has frequently labelled legitimate, factually correct news stories as 'FAKE NEWS/WITCH HUNT!' via Twitter, which leaves many of us wondering what 'truth' is, and if it even exists anymore (on that: if you are up for some mind-bending ideas on truth, head to Chapter 8 after this section). As a result of Donald Trump using the term so frequently with this meaning, it is difficult for people, including students, to know which organisations to trust as reliable sources.

Because of the 'fuzziness' that has arisen around the meaning of the term 'fake news', some governments (like that of the UK) and press associations have decided to stop using it. However, the term is still frequently used on social media and we endure daily bombs of hoax news. In this book, therefore, we will take it at its original meaning: news that is false, and which is generally intended to serve some other purpose, often devious.

If you think about the fake photos and statistics that fuelled the 2019 xenophobic riots in Pretoria and Johannesburg (Mumbere, 2019), or the more strategic and lengthy 'dirty' campaign to stoke racial tensions in South Africa on behalf of Oakbay Investments, a campaign executed by PR company Bell Pottinger in around 2017, you can see how fake news is still a real and a dangerous phenomenon with real-world consequences. Fake news is certainly not isolated to the United States or Russia, so South African students should be just as aware as those who live abroad of the dangers of fake news in society.

Perhaps more importantly, Wendling (2018) argues that it is dangerous to assume that those who are prone to 'falling' for fake news are those with little education. He argues (Ibid.) that highly educated people can also be fooled into believing these stories, with the additional challenge that they can be even more stubborn than less-educated people in clinging to these beliefs when challenged. This is bad news (but not fake news), and means that those in academia, both students and lecturers, need to be especially critical of resources and cautious, lest we perpetuate falsities in our work or social settings.

The popularity of social media seems to fuel the spread of this kind of story – it is so easy to share or comment on a post online, without checking its pedigree, that we often do not even read the contents of articles before doing so. If you have never done this, you can congratulate yourself; but most of us have been guilty of this at some point.

'Fake news vs Real news' by Nicole Kempskie. Copyright © 2016 Dotty's Doodles (https://www.dottysdoodles.com/)

Finally, it is critical to acknowledge that fake news has played an important role in how uncertain we feel about what we know to be true, and what we can label as fake; who we can trust, and who we should permanently block from our feeds. We need to make sure we recognise the truth when we see it, and if it is buried under a mountain of fake news, we need to be prepared to dig for it. Doggedly. Our search for credible information has never been more important.

What do we do about this situation? To start with, we should be using extensively the many fact-checking organisations that exist – a little more on these follows in Section 5.4.3. Wendling (2018) says that organisations like Google and Facebook are trying to avoid being the carriers of misinformation; for this reason, they are hiring extra (human) content reviewers and deploying Artificial Intelligence (or AI) programmes to monitor posts and check user profiles for fake accounts. Some countries like Germany are also implementing legislation that requires social media sites, like Facebook, to remove fake news and hate speech, or risk paying hefty fines.

As for what you can do, a good start is to be aware of fake news and its purposes, and take a critical approach to everything you read and share. The process outlined in Chapter 6 for checking the credibility of your content should help too.

 To practise checking for fake news, go to the Learning Zone to complete activities and explore fact-checking sites.

5.3.7 Cyberbullying

Smartphones can play a very positive role in society, but the prevalence of cyberbullying is one of the negative results of their use (Hardman, 2019). We hope that you have never had to endure this yourself at school (and never been a bully yourself), but it is important to realise that bullying is not limited to the school playground. Cyberbullying is a common concern for adults of all ages, including those at university, even if it looks a little different (Hall, 2019 in Rucker, 2019):

- that provocative, awful troll on your Twitter feed, who seems to like nothing better than flaming you with abuse and threats;
- the 'mystery stalker' who went to great lengths to know Vhudihawe's every move;
- that so-called friend, who outed your secret on Facebook;
- that horrible post on the message board claiming that you are a 'slut';
- that WhatsApp that threatened to show you what a 'real woman' likes;
- that TikTok clip of a student being beaten up that a group of your classmates laughed at;
- or even that student who **catfished** your name and now 'owns' Facebook and Twitter profiles with your name and face.

A **catfish** is someone who pretends to be someone else online by taking another person's identity and/or photo. This is especially common in online dating.

All these are examples of common cyberbullying, although, sadly, they do not come close to covering all the possible forms in which people torment each other online. Hall (2019, in Rucker, 2019) notes that women, some racial groups, and other marginalised people like those from the LGBTQ communities are often targeted by cyberbullying, and the psychological, academic, and even physical consequences of this kind of abuse are significant, extending in some cases to suicide (Ibid.). It is important, therefore, that we recognise the forms cyberbullying can take, and that we know what our rights and responsibilities are as students (and as people!). We offer some practical advice for dealing with cyberbullying in Section 5.4.1 and 5.4.2, but a very good place to start is treating others how you would wish to be treated, and encouraging others to do the same.

Your lecturer can give you a debate activity from the Learning Zone, to help you grapple with the implications of cyberethics.

5.4 Let's get your (digital) citizenship passport stamped!

Phew, that was a lot of heavy thinking, and we are not quite done yet. The next section is far more practical: the tips that follow should help you to navigate your way around the cyberethics issues already discussed, as well as build confidence in your digital citizenship. We will take you through three main pieces of advice in this section – being safe, being good and kind, and being smart. These three elements should help you on the road to digital fluency.

5.4.1 Be safe

Your safety should always be your top concern – here, safety means your physical, psychological, and online security. Earlier, we mentioned how important it is to maintain your digital privacy, and discussed what can happen when you fail to do this, but we also hinted at physical or psychological consequences of problems like cyberbullying. We therefore encourage you to:

Bolster your privacy and safety

Practical ways to boost your privacy, and so protect yourself from physical, psychological, financial, or other kinds of harm, include:

- Checking and managing the privacy settings on all the apps and search engines you use. Where apps allow different levels of privacy on different posts, make sure information that is public is meant to be public. If you are trying to build up a solid reputation for your future, you can make those things that show you in a positive, professional light open to the public. Just make sure that these 'reveals' are an active decision, rather than because you forgot to hide them from prying eyes;
- If you have entered the year of your birth as well as the day and month on social media platforms, deleting it: including the year gives dodgy people easy access to the first 6 digits of your ID number, which is not desirable. Yes, Facebook *will* still remember your birthday even if you just write the day and month, so don't worry, you will still feel all the cyberlove on your special day;
- Shredding documents with personal information on them instead of simply throwing them away or tearing them up – 'one person's trash is another person's treasure' is a true old saying;
- Being careful who you share your personal information with. If you are asked to provide information, make sure the person asking has the authority to do so and a valid reason too;
- Protecting the CVV number on any bank cards you have, and make sure that you keep an eye on your card at all times – under no circumstances allow someone to take your card to a machine out of your sight;
- You are not going to like this one, but it is important: using different passwords for each platform or app, and change them every 90 days – oh, and make sure they aren't too obvious! It helps to work out a code system for this so you can keep track of all of these. For example, choose a single 7- or 8-letter word every 90 days, add in 2 random numbers and a punctuation mark, and then vary one or two letters in the word depending on which app you are using. If you take the word 'elephant' for example, you might use the

password 'W1ephant7!' for WhatsApp, 'T1ephant7!' for TikTok, and ''F1ephant7!' for your FNB banking app. After 90 days, choose a new word, a new set of numbers, and a new punctuation mark and apply the same principle. This appears random and should be more difficult to crack than your birthdate, but it should not be *so* random that you forget them and lock yourself out of your accounts! If this does not work, try a password keeper;

- If you are concerned about passwords and privacy on Google, doing a 'Privacy check-up' and managing your privacy settings from there. Other apps or sites may have similar features;
- If you or someone you know is being harassed or bullied, Hall (2019, in Rucker, 2019) says that not responding or retaliating, but reporting it to your campus security, student services, and/or local police as soon as possible is important – to help the authorities deal with the bully, keep as much evidence about the event/s as possible, including screenshots, emails, URLs or IP addresses, noting the time sequence of these too;
- Taking concrete steps to get the bully out of your life by blocking them, adopting greater privacy measures (perhaps not fair step, but useful), and warning friends and families about the behaviour;
- Finally, if you know the bully or the target of the abuse, doing what you can to stand up to or for them, including speaking with campus or police representatives.

Wipe your feet (or your digital footprint at least)

Have you ever Googled yourself? You might be surprised (and embarrassed) at what pops up. Your digital footprint is the trail of data you leave behind after every visit to the World Wide Web, and so is an integral part of your privacy. If you are not actively managing your footprint, you will have created a 'passive' digital footprint. As we discussed earlier, search engines will identify and organise search results based on this information. Your digital footprint may also influence your credibility and reputation, so be conscious of where you visit and what you do in order to make sure you are leaving the best possible 'footprint'. Consider following these tips:

- Use an avatar to protect your identity – a codename and picture go a long way to protect you from privacy risks on sites like *Instagram* or *TikTok* that do not require your real profile image or name (bonus: avatars and created names can also be an escape hatch for the imagination: RainbowChessBear93 sounds a lot more exciting than Cheryl Siewierski, doesn't it? At the least, if you do use your first name, try to avoid giving away your surname too).
- Keep a list of the online accounts that you have or subscribe to – if you are no longer using one, go to the site and delete your

account, and delete unused apps from your phone. You can manage these from the app menu on your phone, or use a tool like Pocket to keep your accounts under control.

- **TMI** – Although we are sure your friends love hearing about what your 'bae' ate for breakfast, remember that the more information you put out into the online world, the more risk there is to your privacy and security.

> **TMI** is a colloquial abbreviation for too much information – it is used when someone overshares, and tells you something you really don't want to know.

- If you really do need to go into murky online places (we won't ask questions), use a different name and a different email account that you do not mind losing to spam. Depending on which search engine you use, you should also be able to turn on a 'private' or 'incognito' window, which avoids the 'history' function from keeping tabs on where you have been.
- Most websites will alert you to the fact that they use cookies (small bits of software that they put onto your computer) to track you. If you do not want a record that you were there, decline the cookies and get out, or revisit under the cloak of an incognito search.
- If you hate the idea leaving digital crumbs everywhere you go online, you can invest in a Virtual Private Network (VPN) subscription, but these are unfortunately quite expensive. Without a VPN, just remember that any time you actually log on to apps, you are leaving crumbs, so it is worth being cautious about which sites you visit.
- Most importantly, think before you post! Even if you delete a post quickly, someone may still have made a screengrab of it, and it is effectively on the World Wide Web forever.

Choose a different search engine

In Section 5.3.1, we discussed some of the reasons to consider using search engines other than Google. For many people, the value of Google still outweighs the drawbacks and security risks, but you might want to try the alternatives. Priyadarshini (2019) suggests some of the following, depending on what you want to do:

- DuckDuckGo: good for privacy, user-friendly scrolling, and not tracking or bombarding you with adverts;
- Qwant: also emphasises privacy, as well as acting as a music search engine;
- Swisscows: (besides having a cute name) a child-friendly (no porn here!), safety-focused and user-friendly platform for searching;
- SearchEncrypt: good if you would like your browsing history deleted after 15 minutes of inactivity (probably more suited to spies, but an option if you really want to limit your footprint);
- StartPage: footprint-free searches that don't require cookies, customisable and pretty, with different themes available.

Don't catch a virus

Most of us have experienced that dreaded message after a blank screen: infected! Current antivirus software is better at catching bugs and viruses, but we can do more to protect ourselves and our data. Elkholti (2019) recommends taking these actions to ensure that your data and computer stay healthy:

- Use quality antivirus, antimalware, and antispyware software and keep it updated. Obvious! It may feel painful to pay for these (and the data to update them!), but if your computer does not already have a good one installed, it is worth it to avoid losing precious data. If you really cannot afford it, Elkholti (2019) says that free software is better than nothing, even if it does not protect you fully. Try Avast or AVG's free options if you choose this route.
- Keep your operating system up to date (i.e. Windows, Linux or Mac OS) – the updates patch or fix known problems from earlier versions, so the newer your version, the less chance you have of a bug or virus.
- *Very* important: check your Wi-Fi connection, make sure it is protected by a password, and hide your network name if you can. Avoid sharing your password with friends or acquaintances unless you know that their devices are also protected, and you trust them not to give your password to others.
- Avoid using public Wi-Fi – sorry if this breaks your heart! Hackers can access users' information on public Wi-Fi, so try to resist using it.
- Most browsers insist that you use the most up-to-date version, but make sure that they are updated too – like updates to operating systems, security patches for browsers aim to remove previous weaknesses.
- Sharing is *not* caring. If you are downloading 'free' stuff from the World Wide Web, or from flash drives or external hard drives, you may pay the price in the form of spyware.
- Emails can be risky, and phishing scams are sometimes so professionally done that you cannot tell the difference between emails from your real bank and fakes. To avoid this, never click on a link from an email, immediately delete anything that looks like spam, use contact details (websites, phone numbers, etc.) from official documentation, and do not respond to messages that ask you for personal details or your password.
- Finally, if you are confident in how your device works, Elkholti (2019) suggests disabling autorun on Windows to prevent any viruses from breeding like rabbits. However, let us hope that you can keep your device clean and yourself safe without having to resort to more advanced tricks.

Phishing refers to sending a fraudulent email pretending to be from a reputable organisation (like a bank or government agency) with the aim of getting the receiver of the message to reveal personal information, passwords or sensitive financial details.

5.4.2 Be good and kind

There is a great deal of advice about how to be a decent human being on the World Wide Web, but Shea's (2011) core rules still cover the main points about how to stay ethical and be kind online. Table 5. 1 below outlines her 10 core rules.

In addition to the positive actions Shea (2011) outlines, there are also a few things that you should *avoid* doing online while at university. These include:

The colloquial term **noob** refers to someone who is new (a 'newbie') to an environment or inexperienced in a particular activity – it is often applied to people on social media or other online environments like chatrooms.

- SHOUTING THROUGH USE OF ALL CAPS or red pen;
- using offensive language or stereotypes;

Table 5.1. Shea's (2011) Core rules of netiquette

Core rule	In practice …
1. Remember the human	• Practise the golden rule of behaving towards others as you would like them to behave towards you; • Consider your words and actions before you speak or post and ask yourself if you would feel comfortable saying the same thing to someone in person.
2. Follow the same standards of behaviour online as you would in your real life	• Just because you are online does not mean that the rules of society no longer apply; • Act within the laws, rules, and boundaries in the same way you would in real life.
3. Know where you are in cyberspace	• Context matters: your level of formality and the type of content you post and share differs depending on where you are, so keep this in mind.
4. Respect other people's time and bandwidth	• Let's face it, data in South Africa is expensive, and time is precious too, so be concise but clear, and avoid using or downloading bulky files unless they are important.
5. Leave a good impression	• Make sure the parts of you that are visible on social and other media are professional and inspire confidence in your developing personal 'brand'; • Do your 'homework' and research before posting any opinions.
6. Share expert knowledge	• While you might still be learning, share your skills and learning where you can, and make the most of trying to bridge that digital divide; • Give recognition to those whose ideas you are sharing too!
7. Keep your fire extinguisher handy	• Try to avoid online 'flame wars', which, while occasionally amusing to watch, are generally not good for anyone's reputation given the depth of emotion visible in these.
8. Respect people's privacy	• Again, think about the golden rule here – do not share anything you would not want shared, or use anything you would not want someone to use of yours.
9. Don't abuse your power	• With great power, comes great responsibility (and all that ….); • If you are 'at home' on the World Wide Web, make sure that you treat others well, including any noobs; • Help where you can.
10. Forgive mistakes	• We just mentioned 'noobs', but they are important here too: people will make mistakes online, and they may very well annoy you. Try to be patient, and if necessary, be constructive and polite in your corrections.

- overusing emojis or punctuation marks: you know, like these!!!!!!!!!! ☺☺☺☺;
- downloading and sharing copyrighted material – leave the pirating to Jack Sparrow;
- forwarding chain mails, or hitting the 'reply all' button on mails. No. Just no;
- expecting immediate responses from busy people – plan your time and your messages;
- asking questions that you could easily have found the answers to yourself (either through a search engine or reading your prescribed material).

5.4.3 Be smart

You have already made one of the smartest decisions you can make: pursuing your education. In pursuit of becoming a successful student and a lifelong learner, you will need to use the tools that are at your disposal wisely, for example, not falling prey to fake news, or less than credible academic sources.

Chapter 6 and later chapters deal in more detail with checking the credibility of a source and taking a sceptical approach, but in Section 5.3.6 we mentioned fact-checking resources that can help deal with fake news. We suggest you first try Simon Fraser University's Fake News Detection project, which is available at http://fakenews.research.sfu.ca/. This site allows you to paste text into a block, and then do a fact check and genre classification – this identifies the kind of sources the information comes from. For example, if we do a quick test with 'Clinton won the 2016 US election', the site clearly suggests this is a hoax, as the graph shows that no trusted sources make the claim.

Africa Check is another site you can rely on – especially for Africa-related fact-checking purposes. It is available at https://africacheck.org/. This site is a member of the international fact-checking body; it offers to check claims, either online or via email, Twitter, or Facebook, as well as publishing regular reports and fact sheets. They also explain the fact-checking process in detail, so you might find the site particularly helpful for this purpose too.

Finally, we can recommend a visit to Lee Crockett's work at Wabisabi Learning (2019, the URL is included in the reading note in the margin). This is especially useful if you would like to develop your digital literacy and digital fluency. For the same reason, it is worth looking at an outline of the five processes that Crockett (in Wabisabi Learning, 2019) calls future digital fluencies so that you can consider them for different academic tasks:

More on Lee Crockett's work here: https://wabisabilearning.com/blogs/future-fluencies/information-fluency-online-research-strategies

- **solution fluency** helps you to develop problem-solving skills;
- **information fluency** helps you to deal effectively with sources and information, so should help with the 'fake news' problem too;
- **creative fluency** helps you to find artistic flow and focus;

- **media fluency** helps you to develop critical listening skills and make decisions about selecting the best media for your message;
- **collaboration fluency** helps you to develop the ability to work effectively in teams, one of the key soft skills you will need at university and in the workplace.

Wabisabi Learning's site offers some excellent free-to-use resources and tips on how to engage in these five fluencies or processes.

 If you would like to test the abilities of these fact-checking sites, either go to their websites or visit the Learning Zone for some seriously challenging tasks!

5.5 Conclusion

That brings us to the end of what was a chunky chapter full of things to think about and act on. We hope that you are feeling far more comfortable with your status as a citizen of the digital world (big smiles for your Digital Citizenship passport photo please!), and that you feel ready to take on the challenges of working your Google-Fu and finding excellent, credible sources in Chapter 6.

References

Birkinshaw, J. 2018. 'Beyond the information age'. *Wired*. Retrieved 22/08/2019, available at <https://www.wired.com/insights/2014/06/beyond-information-age/>

Buchanan, E.A. and Zimmer, M. 2018. 'Internet research ethics'. *The Stanford encyclopedia of philosophy*, Winter 2018. Zalta, E. N. (Ed.) Retrieved 22/08/2019, available at <https://plato.stanford.edu/cgi-bin/encyclopedia/archinfo .cgi?entry=ethics-internet-research>

Doria, M. 2014. 'If you think we're still in the age of information, think again'. *LinkedIn*, 23 December 2014. Retrieved 22/08/2019, available at <https://www .linkedin.com/pulse/you-think-were-still-age-information-again-mike-doria>

Elkholti, M. 2019. '15 Tips on how to prevent computer viruses in 2019'. *TechCastle*, 6 June 2019. Retrieved 06/09/2019, available at <https://techcastle .com/15-tips-on-how-to-prevent-computer-viruses/>

Eyal, N. 2019. 'How to stay informed without losing your mind', *Nir and Far*. Retrieved 02/09/2019, available at <https://www.nirandfar.com/ how-to-stay-informed-without-losing-your-mind/>.

Friedman, Z. 2019. 'Here are the top 7 websites for free online education'. *Forbes*, 29 May 2019. Retrieved 03/09/2019, available at <https://www.forbes.com/sites/ zackfriedman/2019/05/29/free-online-education/#625eb868342b>

Geach, C. 2014. 'SA pirates a million movies a month'. *IOL*, 14 October 2014. Retrieved 29/08/2019, available at <https://www.iol.co.za/news/sa-pirates-a-million-movies-a-month-1764498>

Google. 2019. *Google privacy policy.* Retrieved 21/07/2019, available at <https://policies.google.com/privacy>

Government of South Africa. 1996. *The constitution of the Republic of South Africa*. South African Government. Retrieved 30/11/2019, available at <https://www.gov.za/documents/constitution-republic-south-africa-1996>

Growth Tribe. 2017. 'The digital skills gap and the future of jobs 2020'. *YouTube*, 4 April 2017. Retrieved 02/03/2019, available at <https://www.youtube.com/watch?v=Y9FOyoS3Fag>

Hardman, J. (Ed.). 2019. *Teaching with information and communication technology (ICT)*. Cape Town: Oxford University Press Southern Africa.

Internet World Stats. 2019. 'The digital divide, ICT, and broadband Internet'. *Internet World Stats*. Retrieved 02/09/2019, available at <https://www.internetworldstats.com/links10.htm>

Kassner, M. 2013. 'Search engine bias: What search results are telling you (and what they're not)'. *TechRepublic*, 23 September 2013. Retrieved 21/08/2019, available at <https://www.techrepublic.com/blog/it-security/search-engine-bias-what-search-results-are-telling-you-and-what-theyre-not/>

Koller, D. 2012. 'What we're learning from online education'. *TEDGlobal*, June 2012. Retrieved 25/08/2019, available at <https://www.ted.com/talks/daphne_koller_what_we_re_learning_from_online_education?language=en#t-634>

Mumbere, D. 2019. 'Fake news fuels xenophobic tensions in South Africa'. *Africa News*, 6 September 2019. Retrieved 07/09/2019, available at <https://www.africanews.com/2019/09/06/fake-news-fuels-xenophobic-tensions-in-south-africa//>

Niselow, T. 2018. 'Five massive data breaches affecting South Africans'. *Fin24*, 19 June 2018. Retrieved 03/09/2019, available at <https://www.fin24.com/Companies/ICT/five-massive-data-breaches-affecting-south-africans-20180619-2>

Priyadarshini, M. 2019. '12 Google alternatives: best search engines to use in 2019'. *Fossbytes*, 21 October 2019. Retrieved 28/11/2019, available at <https://fossbytes.com/google-alternative-best-search-engine/>

Publishers' Association of South Africa. 2019. 'Frequently Asked Questions: Copyright'. *Publishers' Association of South Africa*. Retrieved 02/12/2019, available at <http://www.publishsa.co.za/copyright/faq>

Rainie, L. 2013. 'The state of digital divides (video and slides)'. *Pew Research Center*, 5 November 2013. Retrieved 04/09/2019, available at <https://www.pewinternet.org/2013/11/05/the-state-of-digital-divides-video-slides/>

Ramsey, D. 2019. 'How a data breach can impact you'. *Dave Ramsey*. Retrieved 05/09/2019, available at <https://www.daveramsey.com/blog/data-breach-impacts>

Ribble, M. 2017. 'Nine elements: Nine themes of digital citizenship'. *Digital Citizenship*. Retrieved 22/08/2019, available at <http://www.digitalcitizenship.net/nine-elements.html>

Royal Society for Public Health (RSPH). 2017. '#StatusOfMind: Social media and young people's mental health and wellbeing'. *Royal Society for Public Health*, May 2017 [PDF]. Retrieved 12/11/2019, available at <https://www.rsph.org.uk/uploads/assets/uploaded/d125b27c-0b62-41c5-a2c0155a8887cd01.pdf>

Rucker, M. 2019. 'Cyberbullying on the college campus'. *Affordable Colleges Online*. Retrieved 02/12/2019, available at <https://www.affordablecollegesonline.org/college-resource-center/cyberbullying-awareness/>

Schaefer, J.M. 2011. 'IP infringement online: The dark side of digital'. *World Intellectual Property Organization*, April 2011. Retrieved 30/08/2019, available at <https://www.wipo.int/wipo_magazine/en/2011/02/article_0007.html>

Shea, V. 2011. 'The core rules of netiquette', *Albion*. Retrieved 03/09/2019, available at <http://www.albion.com/netiquette/corerules.html>

South African Depression and Anxiety Group (SADAG). 2019. 'The tips you need to read now about social media and mental health'. *SADAG*. Retrieved 12/11/2019, available at <http://www.sadag.org/index.php?option=com_content&view=article&id=3026:the-tips-you-need-to-read-now-about-social-media-and-mental-health&catid=75&Itemid=132>

Spante, M., Hashemi, S. S., Lundin, M., Algers, A. and Wang, S. (Ed.). 2018. 'Digital competence and digital literacy in higher education research: Systematic review of concept use'. *Cogent education*, 5(1). Retrieved 22/08/2019, available at <https://www.tandfonline.com/doi/full/10.1080/2331186X.2018.1519143>

Van Noorden, R. 2013. 'Open access: The true cost of science publishing'. *Nature International Weekly Journal of Science*, 26 June 2013. Retrieved 02/09/2019, available at <https://www.nature.com/news/open-access-the-true-cost-of-science-publishing-1.12676>

Wabisabi Learning. 2019. 'The future fluencies'. *Wabisabi Learning*. Retrieved 09/12/2019, available at <https://www.wabisabilearning.com/future-fluencies>

Wadhera, M. 2016. 'The information age is over: Welcome to the experience age'. *TechCrunch*, 10 May 2016. Retrieved 22/08/2019, available at <https://techcrunch .com/2016/05/09/the-information-age-is-over-welcome-to-the-experience-age/>

Wendling, M. 2018. 'The (almost) complete history of "fake news"'. *BBC Trending*, 22 January 2018. Retrieved 30/08/2019, available at <https://www.bbc.com/ news/blogs-trending-42724320>

Zimmer, M. 2018. 'Addressing conceptual gaps in big data research ethics: An application of contextual integrity'. *Social Media + Society*, April–June 2018: 1–11.

CHAPTER 6

Work your 'Google-Fu'

'[W]e learn more by looking for the answer to a question and not finding it than we do from learning the answer itself.'
– Lloyd Alexander (2006)

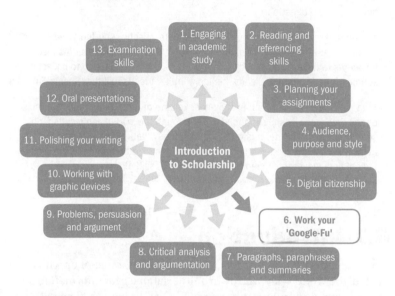

- 1. Engaging in academic study
- 2. Reading and referencing skills
- 3. Planning your assignments
- 4. Audience, purpose and style
- 5. Digital citizenship
- 6. Work your 'Google-Fu'
- 7. Paragraphs, paraphrases and summaries
- 8. Critical analysis and argumentation
- 9. Problems, persuasion and argument
- 10. Working with graphic devices
- 11. Polishing your writing
- 12. Oral presentations
- 13. Examination skills

Introduction to Scholarship

OBJECTIVES

At the end of this chapter, you should be able to:

- explain the value of online search engines and databases;
- understand the differences between the Internet, the World Wide Web, and search engines;
- develop and refine targeted online search phrases for online and database searches;
- use basic Boolean logic to refine online and database searches;
- employ suitable reading techniques to identify potential sources from online search results;
- evaluate the credibility and suitability of sourced information;
- use current online skills as a basis for practising more advanced online search skills.

IRL

Jana, a first-year physics student, has been given an assignment in which she needs to explain how physics can ultimately unify the four identified forces of nature. She has been told to use at least one journal article, and another two credible sources to support her argument.

Unfortunately, after twenty frustrating minutes hacking away at the library database, Jana realises that despite attending the 'how-to' session at the library, she actually does not know how to find the information that will help her answer the question.

As she looks across the room at her classmate Achebe, she feels even more despondent: after just a few minutes of typing and skimming over his results, he seems to have found exactly what he needs, and is getting up to collect his sources from the library. Clearly, he has what her friends call the 'Google-Fu'.

Jana sighs as she thinks about the sheer volume of information about physics on the Web – how is she supposed to sift through the 308 billion results for her search query to find her answer? She absolutely 'gets' Facebook and can flit her way around the Twittersphere like a hummingbird, but apparently, the simple task of finding just three credible sources is impossible!

6.1 Introduction

Jana's experience is not unique – at some point, most of us who have 'surfed' the World Wide Web or an online database for information have experienced that frustrating sense of defeat. In fact, as Reed (2014:21) points out, the word 'surfing' is perfectly suited to what most users are doing: trying to control, and 'artfully use, a force much bigger than they are' without really understanding what is going on underneath them.

So how *do* we 'artfully' control this force? And just how big is it anyway?

Although it is difficult to get exact figures (or even imagine them!), the International Data Corporation (in Patrizio, 2018) estimates that by 2025, worldwide data will reach 175 zettabytes (ZB). If you do not know how big that is, don't feel alone: it's quite unfathomable. Think of it this way: there are already almost 2 billion websites on the World Wide Web (Internet Live Stats, 2019), all of which include content, which translates to data and size. Consider the fact that any Web search you do will search all of these sites, unless you insert parameters – then you may realise just how miraculous it is that you find useful information using search engines like Google at all!

Figure 6.1 (which follows) uses the number of MP3-format songs that you can fit onto a disc of increasing size to give you a mental picture of just how much information is on the Web. To listen to 1 Petabyte of MP3 files, you would need to listen continuously for 2 000 years to get from the first to the last song (McKenna, 2013), so the sheer expanse of the scale of information is evident.

Visit the BBC Future site for a mind-blowing picture of the scale of information on the Internet. This infographic by the Information is Beautiful Studio provides an excellent illustration of these different scales, and reflects the exponential pace at which information is growing. Available at <http://www.bbc.com/future/story/20130621-byte-sized-guide-to-data-storage>

Figure 6.1. Scale of the Internet in MP3 music files

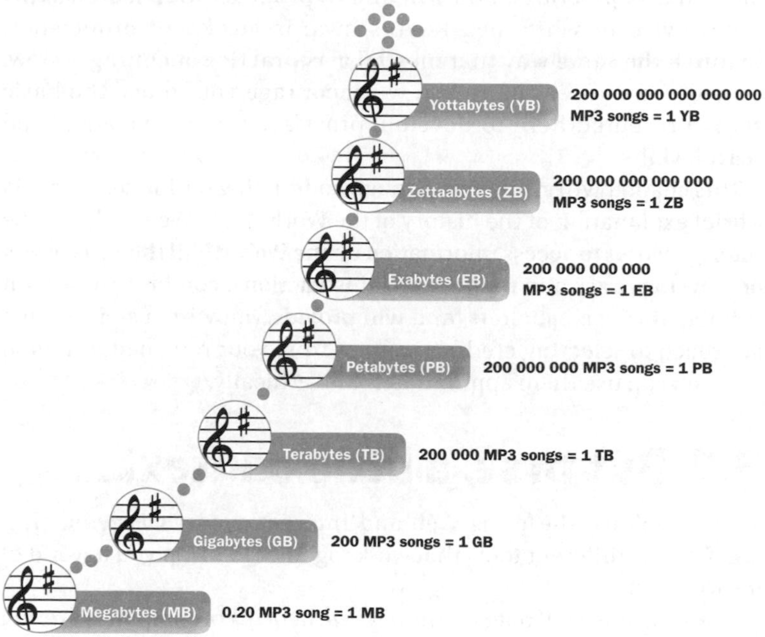

Yottabytes (YB) — 200 000 000 000 000 000 MP3 songs = 1 YB

Zettaabytes (ZB) — 200 000 000 000 000 MP3 songs = 1 ZB

Exabytes (EB) — 200 000 000 000 MP3 songs = 1 EB

Petabytes (PB) — 200 000 000 MP3 songs = 1 PB

Terabytes (TB) — 200 000 MP3 songs = 1 TB

Gigabytes (GB) — 200 MP3 songs = 1 GB

Megabytes (MB) — 0.20 MP3 song = 1 MB

Adapted from BBC (2019) and McKenna (2013)

In your everyday life, you might make use of megabytes, gigabytes, and perhaps even a terabyte or two. You store information, assignments, photos and other media on small flash drives, you upload, download and maybe even create content on the Web, and, if you have a one-terabyte hard drive, you may never worry about running out of space for it all. But, as you can see in Figure 6.1, there are many more levels of scale in the information universe.

The implication of this is important: when you conduct a Google-search, the results that are returned on the first page are like the tip of an iceberg, whose greater mass is hidden underwater (somewhere in those 308 000 000 000 results!). Unfortunately, if you enter an imprecise search phrase, what you are looking for may be hidden underwater; given the size of the Web, it is clearly impractical to go through *all* returned items.

Your ability to create precise, considered search phrases is therefore really important if you want the most relevant items to appear on your first page of returns. The same applies to any search that you might conduct in your university's journal database – if you can use the right words and search phrases, you are much more likely to find the kind of information that will support your arguments convincingly.

You may notice that this chapter purposely moves away from describing your use of the Internet as 'surfing'. Instead of simply trying to glide above the 'beast' that is the World Wide Web with little awareness of what lies beneath it, your new goal is to learn how to master and control the Web through understanding how it works.

In Japanese martial arts, **black-belt** status does not demonstrate mastery and completion – it simply indicates that the martial artist is competent in the basic techniques of the style, and so has achieved the first step towards advanced learning! (World Martial Arts Center, n.d.). This is very much the approach we take in this book: use this book as a platform to learn the basics, and then keep improving once you have mastered what it has to offer!

Boolean operators, named after 19th century mathematician George Boole, refer to three simple words, namely AND, OR and NOT, that are used in different forms and combinations to filter and find information on databases and the Web. For example, 'Achebe advised Jana to use Boolean operators in her search to filter her results.'

If words are used **synonymously**, it means that they are used inter-changeably, as synonyms or, in simpler terms, that they mean the same thing. For example, 'The words "fair" and "just" might be used synonymously in a court of law.'

The **Cold War** refers to a period of military and political hostilities between the United States and its allies, and powers from the Eastern, Soviet bloc between approximately 1947 to the early 1990s. Although little outright, large-scale fighting occurred, both sides used nuclear threats, espionage and technolog-ical competition to gain an advantage over the other. The development of the Internet was just one part of that 'race'.

Point of interest

For a more thorough look into the creation of the Internet and the World Wide Web, visit The Internet Society's visual timeline at https://www.internetsociety.org/timeline/history/.

As a result, in this chapter, we use a metaphor related to various martial arts practices, and will aim to progress your search skills from novice (or 'white-belt') status towards 'black-belt' proficiency. In much the same way that martial arts practice encourages slow, considered mastery of an art, we encourage you to use the basic skills presented here to develop, practise, and evolve advanced search skills.

This chapter will guide you through conducting online searches via a brief explanation of the history of the World Wide Web and how the Internet works to access information on the Web. It will then coach you on how to phrase and refine search instructions, combine these with suitable **Boolean operators**, and will provide you with a step-by-step approach to selecting credible sources from your returned results so that you can use them appropriately and ethically.

6.2 But first, a little context ...

Some people use the terms 'Web' and 'Internet' **synonymously**, but they are actually different tools that work together to bring you a world of information.

The term 'Internet' refers to the *physical network* of computers around the globe that are connected to each other for communication purposes (Techopedia, 2019). Originally developed to overcome potential glitches in military communication during the height of the **Cold War** in the 1960s, this network uses telephone lines, cables, fibre optics and satellites to connect computers, and supports email, newsgroups, instant messaging *and* the World Wide Web.

The World Wide Web on the other hand, refers to a software appli-cation developed by Tim Berners-Lee in 1989 that uses the Internet's physical platform to work. It is a collection of web pages connected through URLs and hyperlinks, so it is just one of the (extremely useful) services provided by the Internet. Techopedia (2019) probably explains the difference between these two concepts most simply: the Internet is the actual hardware and infrastructure of the network itself, while the Web is one of the services offered over the Internet.

A **URL** (or Uniform Resource Locator) is the specific address to a resource on the Internet, while a **hyperlink** is an electronic link to a particular resource or address on the Internet. A hyperlink is activated by clicking on a highlighted or underlined word or image. For example, in an electronic format of this book, the URL http://www.oxford.co.za/ is hyperlinked, and by clicking on the link, you would be able to access the website for Oxford University Press South Africa.

As a concept then, the Internet is larger than the Web alone; it is really the framework on which the Web works. What this means for you is that information stored in computers and servers from all over the world becomes available (to an extent) to you, via the World Wide Web.

The volume, breadth and depth of information available on differ-ent websites gives you a definite advantage over students of previous generations. However, this same blessing also poses a new problem – the one that Jana experienced earlier in this chapter: just how do you find the *right* information? Fortunately, search engines like Google or Bing exist to help you search and find information quickly and effectively, as we mentioned in Chapter 5, instead of having to manually locate and read through all the websites available to you.

In brief, **search engines** are Web-based tools that use software applications like 'spiders' and 'bots' to follow links to all websites that include the search words or phrases that you have entered. Once a search engine has located these links, it reports all the relevant websites to you in the form of ranked results, from most to least relevant (hopefully). The functions of a search engine, according to SEM Advisory (2017), are therefore to:

- **crawl** around the Web using spiders or bots to collect as much information about existing, new or updated websites as possible;
- **index** keywords and other information found into a massive catalogue of all relevant words, tags, and content;
- **store** information (collected by the bots and spiders) in its database to speed up searches; and finally,
- **provide results** to specific keyword searches by matching these to its crawled, stored, and indexed pages – different search engines will produce different results because they use different **algorithms**.

Google, or 'Professor Google' as it is sometimes affectionately called, is still the world's leading search engine (Statista, 2019), and helps users to sieve through enormous volumes of information and formats in a matter of seconds. In fact, on average, Google processes almost 78 000 searches from around the world every second! (Internet Live Stats, 2019). If you learn how to use search engines like Google effectively, you will save a significant amount of time in searching for information, and also ensure that you source the most relevant information for your purposes.

> Search engine algorithms are formulas or rules designed to help organise, filter, identify, and rank content on the Web. For example, algorithms like Panda help to filter and judge content, while Penguin's job is to identify spam! To read more about this, visit Dave Davies' 2018 article at <https://www.searchenginejournal.com/how-search-algorithms-work/252301/#close>

Point of interest

For a visual depiction of this and other mind-boggling one-second usage statistics that include Twitter, Instagram, and others, visit Internet Live Stats at <https://www.internetlivestats.com/one-second/> and click on the 'scroll' button for maximum effect.

6.3 White-belt 'Google-Fu'

Right! By now you should have a pretty good idea of what the Internet, the Web, and search engines are, and from Chapter 5, you should also understand some of the possible obstacles and ethical difficulties you might encounter. But, like Jana, you might not yet know how to actually go about *finding* your information. It sounds easy enough, right? You open Google, then type a few words and bam – there's your answer!

Um, not quite. There is an art to using search engines to find information, and, as with any martial art, practise makes perfect. The next section outlines the techniques that will help you to develop your 'Google-Fu'.

> **PRACTISE NOTE**
>
> You may certainly use other search engines or databases for your research, but, for the purposes of demonstrating this technique, we will make use of Google, which remains the most commonly used general search engine (Statista, 2019) despite some concerns around its practices. For this particular exercise, we will use Google Scholar (https://scholar.google.com/), which prioritises journal articles and academic content, to complete our practise search. Ready? Let's go ...

6.3.1 Developing search phrases

In Chapter 3, we took you through the steps of analysing your question. The skills that you learnt there are going to be useful here too, since identifying the main, or key, word is one of the most important elements of topic analysis.

To start, let us say that you have been asked to *detail the effects of labour strikes in the mining sector of South Africa between 2010 and 2013*, and to cite at least two academic articles, as part of an assignment.

We can use a three-step approach to finding the relevant phrase or phrases:

- **Step 1: Underline** the words in the instruction that you feel most reflect what is required (for now, don't worry about the instruction verb). Something like this:

Detail the effects of the labour strikes in the mining sector of South Africa between 2010 and 2013.

Point of interest

You may notice that Google Scholar offers autocomplete suggestions as you are inserting your search phrases. For most academic searches, you will need to ignore these options and continue typing your own terms, but you may occasionally find that these are useful.

- **Step 2:** From the underlined words, **identify a broad keyword** that is relevant to your search. This will help your search engine or database to restrict its search to a particular field:

In this case, the word 'strikes', or even 'labour strikes' will work.

- **Step 3:** Now **identify narrow**, or **delimiting, words or phrases** that will help to refine your search. For example, this will help your search engine to eliminate articles that include the words 'labour strikes', but do not include 'mining sector', 'South Africa', or the dates in question. Again, this will help you filter out a lot of unnecessary content:

In this case, the phrases 'mining sector', 'South Africa', and '2010 to 2013' will narrow our search so that our returns are more specific, and not about labour strikes in general.

Great! We have now identified particular words and phrases that may help us find the right kind of text. Let's try it out on *Google Scholar* and see what happens.

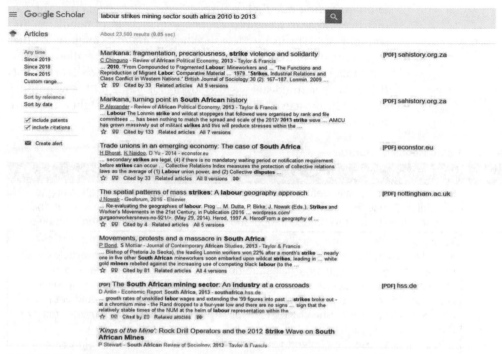

Google and the Google logo are registered trademarks of Google LLC, used with permission.

Let's quickly skim read the headings and brief descriptions of each result, starting at the top and moving down the page.

These results are promising, but they are not exactly what we were looking for. There are a whopping 23 500 results! Despite including '2010 to 2013' in the search keywords, we can also see some results from 2014 and 2016. These might deal with strikes that occurred between 2010 and 2013, but we cannot be sure without checking. The prioritised returns are also more focused on Marikana and strikes in general – not necessarily on the effects of mining strikes because we did not specify 'effects'. Some of the results are from journals you need to pay to access, which might also be problematic – for example, the Marikana article by Alexander costs $43 and we don't even know if it is relevant yet! Happily, there are some potentially useful articles that appear at the bottom of the returned suggestions, and these do not appear to be paywalled, so we will keep these in mind.

For now, we need to refine our search further. The easiest way to do this is to start using Boolean operators.

Paywalled content is hidden from view or download until the reader has either subscribed or purchased it for a fee.

6.3.2 Using basic Boolean logic with search phrases

As we explained earlier in Section 6.2, search engines make use of logic to organise and rate their returned items. This means that you can instruct your search engine or database to include, or remove, or find similar results, by using a few specific commands. Google and other search engines and databases use punctuation to perform these actions.

You can find a tilde ~ on a regular (QWERTY) keyboard by pressing the Function key and the key to the left of the number 1, or by pressing SHIFT and the key to the left of the number 1. It is often below the ESC (or Escape) key.

For this, we use Boolean operators. These are certain words or symbols that help us to find what we are looking for. Table 6.1 below outlines some of the most useful Boolean operators that you might want to use to refine your search.

Table 6.1. Basic Boolean operators

Symbol	How to use it
" "	Search engines usually treat each word as a separate item. If you are looking for a specific phrase and do not want your search engine or database to split up the phrase, place the entire phrase in quotation marks. For example, "labour market".
-	To exclude words or particular items, simply add a '-' sign directly in front of the item. For example, -*Taylor & Francis* in our earlier search will remove any articles from the publishing house – their articles are not useful to us because they require extra payment. This operator is also useful when one word refers to many things. For example, in Jana's case, if she typed in 'forces of nature', she would get many references to a movie. She could then type in "forces of nature" –movie to remove these irrelevant items. You can also use the '-' sign to connect two or more words though. In this case, the search engine assumes that you are looking for words that are strongly connected. This is not as specific as using quotation marks around a phrase, but it does give the search engine an instruction to look for these words together. For example, eighty-year-old patient.
*	Use an asterisk inside a search as a 'wildcard' if you do not know all of the words. This is useful when you are looking for a particular expression or quote but don't know all its details. For example, you would type "a * in time saves *" to find the full expression 'a stitch in time saves nine'.
..	If you separate two numbers with two period stops and no spaces using Google, you are instructing the search engine or database to use these numbers as a range within which to look. For example, if we used *1899..1903*, we would cover the years from 1899 to 1903 in our search. Google Scholar now has a customisable range in its margin so this operator no longer works on Google Scholar. However, it does still work on Google.
OR	If you would like the search engine or directory to search for options that are unlikely to appear together, you can insert the word OR next to the item. For example, if you wanted the search to return items from any year in a three- or four-year period, inserting all the years may not produce any results, since the search engine would look for all of the years. Adding in *OR* after each year would allow the search engine to find options from any one of those years. For example: 2010 OR 2011 OR 2012 OR 2013
~	If you are looking for related words to a topic, you can insert this sign (called a tilde) next to the word you enter into your search. For example, if you are looking up information on university, you might type in ~university, and get results relating to higher education and colleges too.

Okay, let's see what happens when we add in the keyword 'effects' as well as some Boolean search operators to our example search:

- Firstly, we add quotation marks to "labour strikes", "mining sector" and "South Africa" so that our search engine knows it needs to treat these words as phrases, and not as individual words.
- Secondly, we add in an OR operator for information on any of the years we are looking for (i.e. 2010 or 2011 or 2012 or 2013) – we could also just use the new custom range that Google Scholar offers in its left margin but for this exercise, we will go with the Boolean operator.
- And thirdly, we restrict the search to exclude any articles from Taylor & Francis by doing this: –Taylor & Francis.

Let's give it a bash ...

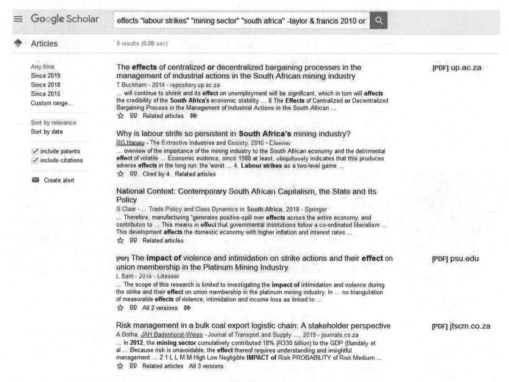

Google and the Google logo are registered trademarks of Google Inc., used with permission.

Okay, let's scan through the fresh results. They do look more promising than the first results. There are only eight returned results, at least two of which (the ones by Buckham and Sam) seem to relate to the effects of strike action and labour bargaining in South Africa. We still have a few paywalled articles (Elsevier and Springer), but at least the Taylor & Francis one is gone. What now?

6.3.3 Using reading skills to filter and evaluate the credibility of returned results

An **abstract** is a brief summary of a research article, thesis, or conference proceeding that appears at the start of the written piece. It outlines the main argument and findings of the author or authors, and can be anywhere from a single paragraph, to an entire page in length. For example, 'Achebe read through the article's abstract to get a better idea of what it was about.'

A really useful academic habit for this stage of your online searching is to record the full **reference** for any items or articles that might be useful on a separate document. Next to your entry, write a few key words about what the article covers, so that you know where to find which information later.

Note!

As mentioned earlier, not all returned results will be free to access. You might be asked to pay for access to some journal articles unless you are using your institution's database (and even then, you may be asked to pay!). As a general rule, you should have access to more articles and databases than you will ever need if you use your library's network to access them. Since your institution pays licence fees to journals, you will not need to pay extra, as long as the journal you have sourced is part of your institution's collection.

The descriptive detail on the results page provides you with important information about the sources. For example, it tells you how many other people have cited the source, in which year it was published, whether there are any related articles, the number of versions that appear, and it also provides you with a brief description of the content by including your search phrases in bold.

You can explore any of the elements of the brief description that are underlined, and Google Scholar will link you to the relevant citation, source or version.

Once you have found a source that you think might be suitable, click on the link to open it. You are likely to find an abstract.

Using your skimming technique, read the abstract to see if any of the main points of your topic are discussed in the article. If they are, then it is very likely that the article may be relevant to you.

If you do not see anything about your particular topic in the abstract, there is still a chance that your topic is discussed in the full article, but it does make the source less likely to be a good one. If you think the article might still be useful for extra detail, make a note of the **source**, but then move to the next most likely one.

If you have immediate access to the journal article or full text itself, using a CTRL + F search might save you time. By pressing and holding CTRL then pressing F, you open up a search bar in which you can type the particular words or phrases that you would like it to look for. It will then take you directly to the location/s of those words or phrases in the document, one by one. Working this way means that you do not need to examine each potentially useful article in detail right away – instead, you can leave the critical reading for when you are actually engaging with your texts.

You should repeat this stage until you have checked all of the likely returned items. For our earlier search, at least three returned items look highly relevant, as they all clearly related to strike action in the mining sector, and specifically included words relating to the costs and effects of strikes in that sector.

A few of the returned options do not appear to focus specifically on the effects of labour strikes in South Africa, although they may include some relevant information. If the promising ones do not turn out to be credible, we may need to do another search. But first, let's see if we can use them.

For now, try to locate the full text of at least three of the most promising items. In this case, fortunately for us, the three most relevant options of the eight items are free access documents, and available as PDFs, so we are able to access the full text even if we are not using our library's network.

Once you have had a chance to skim through the full texts and have confirmed that they are suitable, it is worthwhile to check whether these sources are credible before you start to design any arguments around them. The Web is full of ideas that are best left out of your argument, so be critical and vigilant. We dealt a little with checking facts and credibility for general purposes in Chapter 5, but a more rigorous approach is necessary for academic sources. You can check for credibility by:

- **Understanding and recognising what academic writing entails**. If your source is not using an academic writing style and referencing ideas, move on – quickly!
- **Checking to see if a journal is peer-reviewed**. In general, peer-reviewed academic journals are more credible than professional journals, or textbooks, and are certainly more credible than news reports or magazines. Simply type the journal name into your search engine, and visit the 'About' section of the journal's website to find out more about it. Depending on what kind of research you are doing, however, you may need to use any, or even all, of these sources. Just understand that the further you move away from academic journals, the less scientifically rigorous, and more opinion-based, the writing is likely to become. For example, the second result from our earlier search, by Harvey, is from a double-blind peer reviewed Elsevier journal called *The Extractive Industries and Society*, so we can be confident that this article is credible.
- **Checking the scholarly reputation of an author** – you can do this by examining the number of citations next to the article (the more the better), by looking them up on WorldCat, Web of Science or on another database or search engine. If they are a better-known author, you can even do a quick check on Wikipedia, since any scandals or criticisms are likely to be mentioned there. You can also check on authors of textbooks or news articles in the same way, depending how popular they are. Be wary of social media 'hits' though – a LinkedIn profile can make grand claims, but these claims are not verified so they might not be true. Find out what the author's educational background is, and what experience they have in the area they have written about.
- **Looking for the author's contact information and educational or professional affiliations**. For example, in the sources we identified earlier, one of the articles is from the University of Pretoria's academic publication repository, so we can be relatively sure that we can trust it as a source, even if it has not (yet) been cited by others.
- **If the information is online, checking the domain name for clues**. For online sources, what you ideally want to see is **.edu**, **.ac**, or **.gov**, in the domain name, and not **.org** or **.com**, which

is used for organisations and commercial sites. Using the first three domain types is not guaranteed to eliminate 'fluff', but anyone trying to 'sell' you information from a site ending in .com may easily have a motive other than education. Non-governmental organisations (using .org domains) can also vary from exceptionally credible to really, really bad. Be wary of these.

- Check the sources of your source. If your source has cited other credible sources, it is much more likely that your source is credible too.
- Ask yourself these questions:
  - > Is the website educational and provided by a credible organisation?
  - > Is it trying to sell me something?
  - > Can I detect bias in the author's views?
  - > Is the research paid for? Payment does not automatically discredit a source, but, for example, a study funded by a liquor company that finds that drinking is not as harmful as we thought is definitely suspect.
  - > Is it current? (For the most part, try to avoid publications older than five years unless you are specifically told not to, or you have to refer to literary, historical, legal or primary sources – these are generally acceptable even if they are 'old'.)
  - > Is it published by a reputable organisation? (Run for the hills if something is self-published.)

6.3.4 Refining your search

In our earlier example, we managed to find two really good, and very suitable matches for our topic: Buckham, 2014; and Sam, 2014. However, we might benefit from another example, either because we want to demonstrate to our lecturer that we can do more than meet requirements, or because we think that we need some extra background information.

If this is the case, you can expand or reduce your search phrase using some final tweaks, depending on whether you had too many, or too few results, and also on how relevant the content was.

Because our results were relevant but limited to eight items, we may want to look for extra sources. So, we need to remove some of the restrictions on the search. Instead of using the phrase "mining sector", we are just going to use *mining*, removing the word "sector". Instead of "labour strikes", we are just going to use *strikes*, removing the word "labour". These additional elements were limiting the search considerably. The rest of the search can stay the same, since it brought up the right kind of information.

Let's see what that brings us.

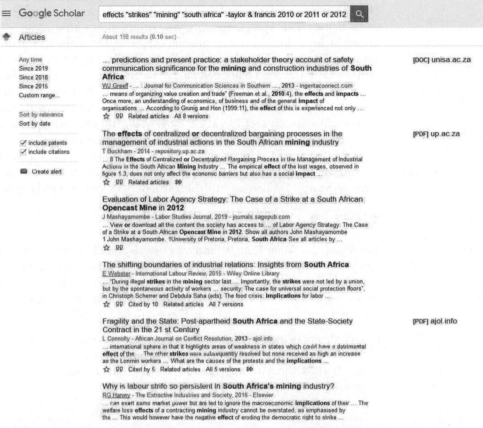

Google and the Google logo are registered trademarks of Google Inc., used with permission.

As you can see, we now have 198 results, some of which we have already viewed. We also have a number of other mining-related items that could be useful. For our purposes, the case study by Mashayamombe of a 2012 opencast mine strike may prove valuable in adding some anecdotal evidence, so this may be the best place to start checking on our options for extra information.

If we repeat the reading and credibility steps with the promising items here, we would probably find at least a few that we could use to buffer and build our arguments. Well done! You've successfully executed an online search.

6.4 Black-belt 'Google-Fu'

Phew, that was a bit of a workout! If you still have energy after all that white-belt searching practice, maybe you will appreciate a few advanced, black-belt level skills. As we mentioned earlier in this chapter, in many Japanese martial arts, the black belt is divided into different levels of achievement. Having a black belt *does not mean that a practitioner has completed training*, but that they have achieved the first step towards advanced learning. The advanced training is then divided into different degrees. This chapter sees 'black belt Google-Fu' in the same way: because the Internet is an organic, constantly shifting thing, we can never know everything about it. But if you have attained black-belt status, you are well on your way to continuing your learning independently and discovering some of the weird and wonderful things on Google, such as 'Easter eggs'.

6.4.1 Easter eggs: a bit of fun

An **Easter egg**, in the sense that it is used here, refers to a hidden message or surprise that is built into the software by software developers as a prank. For example, 'That Google Chuck Norris Easter egg was hilarious!'

The teams that keep Google (and other software) alive often seem to enjoy embedding tricks and 'Easter eggs' into their software. If you are looking for a little light-hearted entertainment as part of your black-belt experimentation, play around with some of these classic online Easter eggs (naturally, your first black-belt task is to find them!):

- Google Pacman, Tic Tac Toe, Minesweeper and Atari Breakout (yes, you can actually play games on Google);
- the Thanos Effect (useful if you would like half of your search results to disappear!);
- Google flight simulator;
- Fidget spinner;
- Flip a coin;
- Google 'Do a barrel roll' trick;
- the very cool LGBTQ (Pride) egg for Google Docs Spreadsheets;
- Google's LMGTFY is a cheeky one, but still fun;
- the Konami Code (↑↑↓↓←→←→ba), probably well-known with gamers, is another one that companies often use on their websites to hide Easter eggs;
- Zerg Rush;
- Windows 10's Emoji Keyboard (fun for social online activities, but not too useful for academia);
- the answer to life, the universe, and everything.

6.4.2 Advanced tips for searching

Okay, we are almost at the end of the chapter, and yet, like any good black-belt candidate, you still feel that you could use a few more practical tips to improve your online search habits. This section gives you a list of really useful, more advanced suggestions (adapted from Verma, 2015 and Catone, 2011) that you can use to further refine your searches and really prove your Google-Fu to the world.

Black-belt function	How to use it
Try different search engines	For a variety of reasons (as discussed in Chapter 5), you might like to use a search engine other than Google. Try DuckDuckGo for enhanced privacy, Swisscows for child-friendly searches, Search Encrypt if you want your local browsing history deleted every 15 minutes; or StartPage if you want a pretty theme to explore websites from – all without using (trackable) cookies. If you are really into privacy and locating uncensored content, you might give Gibiru a try!
Restrict a search to just one site (site:)	Select the site you want to source an item from, and type *site:* in front of it. For example, site:dailymaverick.co.za
Find only one type of file (filetype:)	To exclude other filetypes, simply type *filetype:* in front of the file format you are looking for. For example, filetype:MP3 will find only MP3 files
Restrict a search to words in a title (intitle:)	If you want to make sure that a search return adequately covers your topic, you can do a search for the word in titles only by adding in *intitle:* in front of the word you are looking for. For example, intitle: "labour brokers". This will restrict the search to any titles that have the term labour brokers in them.
Restrict a search to a particular author (author:)	To find a particular author and avoid any other mentions of a name, use *author:* at the start of your search, and follow it by the subject. For example, author:Meredith histology slides. This will ensure the search ignores any mentions of 'Meredith' that are unrelated to histology slides.
To find a definition (define:)	Type *define:* to find a definition or explanation of a word. For example, define:melodramatic
To convert units	This is really simple on Google – simply type a sentence that includes the units and amounts. For example, 160 000 km in meters
Calculator	For simple mathematical problems, you can type out your query using the normal mathematical symbols, such as * [multiply], / [divide], + [add], − [subtract], = [equals] and ( ) [brackets]. You do not have to type 'calculate'. For example, entering: 93/2*654= will bring up an image of a calculator and the answer to your query.
To locate an expired URL	There is almost nothing more frustrating than clicking on a link that goes nowhere. If a link is broken, open the Internet Archive's website at archive.org and simply enter the URL. It will return a calendar with dots on all the dates that your URL was last 'recorded' – click on the one closest to your search requirements and – hey presto! You should have your URL open.
Trying to copy the address of a link without opening it	We all know how irritating this is – it happens often when you are dealing with your sources. When you are mousing around, you can avoid opening a link by holding down the 'Alt' key on Windows computers, or the 'Option' key on Macs.
Locate similar websites	You can access similar kinds of websites by visiting AlternativeTo.net

Black-belt function	How to use it
Get back accidentally closed tabs	It happens – we all have moments when we accidentally close our open tabs and destroy our search. To get your most recent tabs open again, try CTRL+shift+T for Windows computers, or Command+shift+T for Macs.
Be quiet!	For those irritating sites that start auto-playing loud content, you can simply right click on the tab and select 'mute site' – it should be blissfully quiet now.
Intitle search	If you add *intitle:* before your search keywords, the results that come back will refer only to pages with that phrase in the title, increasing the chances of their relevance to your search.
Change your word form	If you are struggling to find useful suggestions from a search, try changing your keywords a little – for example, instead of looking for 'technologies', look for 'technology', and change past or present participle verbs into their infinitive forms (e.g. instead of 'published', try 'publish').
'Speedwatch' on YouTube	If you have to watch a long but useful video on YouTube (very useful for checking if your sources are relevant and credible), you can change the speed setting to 2 – you should still understand it at this faster speed, but you can slow it down for important points, if necessary.
Checking for social tags	If you are looking for mentions of a particular person, organisation or trending topic in a search, simply add '@' or '#' before your search term, even if you're not expecting your results to come exclusively from Twitter.
Frustrated and can't find what you're looking for?	Calling time on a search that doesn't seem to be getting any results is one of the most helpful suggestions. Your time has value, so set a reasonable time limit for yourself; if you reach the limit without finding what you are looking for, change search engines, start a different task and come back to this one, or ask a lecturer, friend or librarian for assistance.

 If you would like to test your black-belt progress, head to the Learning Zone for some seriously challenging tasks! See student exercise LZ 6.2.

6.5 Conclusion

That brings us to the end of what we hope was a practical, useful chapter. Progressing from white-belt to black-belt status (and beyond) is not difficult – it just takes practice, a little logic, and some understanding of how search engines work. So keep exercising your Google-Fu! In this chapter, we covered the basics behind the Internet, the World Wide Web, and search engines, and then focused on executing online searches using search engines or databases. If you follow the principles explored in this chapter and practise your skills frequently, you should have no problem developing some serious Google-Fu for your studies!

References

Alexander, L. 2006. *The book of three (The chronicles of Prydain, book I)*. Chapter 1 [Revised Kindle Edition]. New York, NY: Holt.

BBC. 2019. 'Byte-sized graphic guide to data storage'. *BBC Future*, feature by Information is Beautiful Studio, 21 June 2013. Retrieved 27/08/2019, available at <http://www.bbc.com/future/story/20130621-byte-sized-guide-to-data-storage>

Catone, J. 2011. 'How to use Google search more effectively [infographic]'. *Mashable*, 24 November 2011. Retrieved 29/08/2019, available at <http://mashable.com/2011/11/24/google-search-infographic>

Davies, D. 2018. 'How search engine algorithms work: everything you need to know'. *Search Engine Journal*, 10 May 2018. Retrieved 30/08/2019, available at <https://www.searchenginejournal.com/how-search-algorithms-work/252301/#close>

Internet Live Stats. 2019. 'Internet Traffic on Google in 1 Second', *Internet Live Stats*. Retrieved 30/08/2019, available at <https://www.internetlivestats.com/one-second/#google-band>

McKenna, B. 2013. 'What does a petabyte look like?' *Computer Weekly*, 20 March 2013. Retrieved 30/08/2019, available at <http://www.computerweekly.com/feature/What-does-a-petabyte-look-like>

Patrizio, A. 2018. 'IDC: Expect 175 zettabytes of data worldwide by 2025'. *Network World*, 3 December 2018. Retrieved 29/08/2019, available at <https://www.networkworld.com/article/3325397/idc-expect-175-zettabytes-of-data-worldwide-by-2025.html>

Reed, T. V. 2014. *Digitized lives: Culture, power and social change in the Internet era*. New York, NY: Taylor and Francis.

SEM Advisory. 2017. 'Four Functions of Internet Search Engines'. *SEM Advisory*, 20 March 2017. Retrieved 30/08/2019, available at <http://semadvisory.com/4-functions-of-internet-search-engines-to-help-you-understand-their-importance/>

Statista. 2019. 'Worldwide desktop market share of leading search engines from January 2010 to July 2019'. by J. Clement. *Statista*, 3 December 2019. Retrieved 15/01/2020, available at <https://www.statista.com/statistics/216573/worldwide-market-share-of-search-engines/>

Techopedia. 2019. 'Internet'. *Techopedia*. Retrieved 29/08/2019, available at <https://www.techopedia.com/definition/2419/internet>

The Internet Society. 2019. 'Brief history of the Internet'. *Internet Society.* Retrieved 29/08/2019, available at <https://www.internetsociety.org/history/>

Verma, A. 2015. 'Top 20 best Google search tips and tricks that you must know'. *Fossbytes*, 3 July 2015. Retrieved 29/08/2019, available at <https://fossbytes.com/top-best-google-search-tips-tricks-must-know/>

World Martial Arts Center. n.d. 'Purpose of the Belts'. *World Martial Arts Center.* Retrieved 28/08/2019, available at <http://www.wmacenter.com/page/purpose-of-the-belts>

CHAPTER 7

Paragraphs, paraphrases and summaries

'If you have built castles in the air, your work need not be lost; that is where they should be. Now put the foundations under them.'
– Henry David Thoreau (1854: Chapter 18)

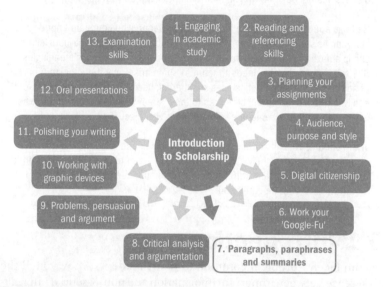

13. Examination skills

1. Engaging in academic study

2. Reading and referencing skills

12. Oral presentations

3. Planning your assignments

11. Polishing your writing

Introduction to Scholarship

4. Audience, purpose and style

10. Working with graphic devices

5. Digital citizenship

9. Problems, persuasion and argument

6. Work your 'Google-Fu'

8. Critical analysis and argumentation

7. **Paragraphs, paraphrases and summaries**

 OBJECTIVES

At the end of this chapter, you should be able to:

· understand the purpose of paragraphs in academic writing contexts;
· identify the main and supporting points in a passage;
· write structurally sound paragraphs;
· use linking words and transitions to logically connect sentences and paragraphs;
· recognise the reasons for paraphrasing and summarising in academic contexts;
· distinguish between a paraphrase and a summary;
· use the paraphrasing method to represent another author's ideas;
· use the method for writing a summary to represent another author's ideas in précis form.

IRL

Since his school days, Nazim has been a diligent student, taking care to record everything his teachers said about the work. He prides himself on doing well academically; taking extensive notes in class definitely helped him to achieve at school.

But after a few months of university, Nazim has realised that his note-taking methods are not working as well at university as they did at school. His lecturers race through the work so quickly that it is impossible to keep up with what they are saying, and they do not even cover everything that needs to be studied! When he looks at his piles of notes from the last two months, he realises that he is in trouble. Aside from the big gaps, he also has little recollection of class discussions because he was so busy trying to keep up with his notes.

He talks about his problem to his older sister Sadaf, who completed her degree last year. As she looks at Nazim's big stack of notes from his last one-hour lecture, she is amazed at how much he has written, and how much of it is meaningless. She tells Nazim that he needs to learn to summarise only the *important* parts of any class discussions. If he uses summaries, he will be able to keep up more easily, and because he will be forced to *think* about what is important and what is not, he might also remember more of what was discussed in class.

When Sadaf leaves his room, Nazim sighs to himself and wonders if he has bitten off more than he can chew – it's easy for his big sister to give him advice about only including 'important stuff', but how is he supposed to recognise what is important and what he can ignore?

7.1 Introduction

Nazim faces a problem common to many students – how to tell the difference between important information and non-essential information, and then record just the important bits. Given the huge amount of content that tertiary students are expected to get through, being able to identify the key points, back them up with evidence, and ignore irrelevant 'fluff', is a key skill that students need to develop.

The previous chapters of this book focused on reading, comprehension, sourcing information, tackling academic assignments, and selecting suitable language for academic tasks. So far, we have not engaged with the writing itself in any detail. This chapter will work on developing your writing skills by building your understanding and practice of writing paragraphs, summaries and paraphrases, all of which are vital activities in your study practices.

As with most skills we aim to develop in this book, remember that the extent to which you practise these writing skills is likely to determine how good you get at writing. For this reason, there are several very useful tasks and activities for this chapter on the Learning Zone.

Importantly, if you apply the skills you learn in this chapter to your other modules, you will reinforce the skills you have learnt, as well as saving time and improving the quality of your understanding of the subject matter.

7.2 Writing paragraphs

At this stage of your education, you can no doubt recognise a paragraph when you see one: paragraphs are essential components of essays and of most other forms of academic response that you will need to write. Despite being able to recognise paragraphs, you might not remember what they are used for. To demonstrate why paragraphs are so important, let us take a look at a sample of academic writing without them. Read the following extract, taken from a university paper, and try to grasp the points made in the extract.

Creecy's (2012) call to raise awareness of cyberbullying, and her anecdotal citing of two suicides is — if superficial — evidence of the potentially lethal effects of cyberbullying. It does not, however, speak to the extent of the phenomenon. Indeed, Creecy (2013) openly admits that an enormous dark figure exists when it comes to establishing the number of children at Gauteng schools who have suffered some form of cyberbullying. Few official studies have been conducted to obtain realistic figures of the phenomenon of cyberbullying in South Africa, since it is, by its nature, problematic in terms of reporting. Few victims wish to acknowledge the humiliation they have experienced (Burton, 2012). A study conducted by The Centre for Justice and Crime Prevention's (CJCP's) Burton and Mutongwizo (2009) however, shows that one in three of all South African children who were interviewed have experienced cyberbullying at school in one form or another. A similar study conducted in Gauteng schools by UNISA in 2012 (Mawson, 2013) supports the figures in the CJCP study, indicating that out of 3 000 students, more than 34% had been victims of bullying over the past two years, with 17% of the bullying being perpetrated via the Internet (Ibid.). This study found that the majority of cyberbullying was conducted using cell phones, and included sexual remarks, threats, false statements, name calling, and 'upsetting messages'. Given the ubiquity of cell phones in South Africa, this finding is disturbing but perhaps unsurprising. What was perhaps a more disturbing finding in the UNISA study, however, was that over 40% of the respondents indicated that they were 'against reporting bullying', either because they were afraid of getting into trouble, or because they were afraid of further victimisation.

Adapted from Siewierski, C. 2013

Did you find that text easy to read? Or did you struggle to hold on to the thread of the argument? If you are like most readers, you will have battled to follow the logic of the argument or to identify the main points. You may even have become 'lost' or bored after a couple of lines and stopped reading altogether! (Don't worry, this is normal – the lack of space between the main ideas makes the piece difficult to understand and absorb.)

If you think back to Chapter 4, you may recall that using white space in your layout helps to improve the readability of a piece of text. This is exactly what paragraphs do: they help structure your ideas by leaving a noticeable white space between one main idea and its evidence, and the next main idea and its evidence. This lets your brain naturally partition the writer's argument into more manageable 'chunks' of information so that it is easier to follow.

The **'dark figure'** of crime is a statistical term that describes the gap between the number of crimes actually committed, and the number of crimes that are reported. For example, 'Car thefts have a low dark figure because people have to report theft to the police in order to claim from their insurance companies. Illegal drug use, however, has a high dark figure, since those who use drugs are not likely to report themselves.'

Let us see if you find it easier to read the same text but with paragraph divisions:

Creecy's (2012) call to raise awareness of cyberbullying, and her anecdotal citing of two suicides is — if superficial — evidence of the potentially lethal effects of cyberbullying. It does not, however, speak to the extent of the phenomenon. Indeed, Creecy (2013) openly admits that an enormous dark figure exists when it comes to establishing the number of children at Gauteng schools who have suffered some form of cyberbullying.

Few official studies have been conducted to obtain realistic figures of the phenomenon of cyberbullying in South Africa, since it is, by its nature, problematic in terms of reporting. Few victims wish to acknowledge the humiliation they have experienced (Burton, 2012). A study conducted by The Centre for Justice and Crime Prevention's (CJCP's) Burton and Mutongwizo (2009) however, shows that one in three of every South African children who were interviewed have experienced cyberbullying in some form or another at school.

A similar study conducted in Gauteng schools by UNISA in 2012 (Mawson, 2013) supports the figures in the CJCP study, indicating that out of the 3 000 students, more than 34% had been victims of bullying over the past two years, with 17% of the bullying being perpetrated via the Internet (Ibid.). This study found that the majority of cyberbullying was conducted using cell phones, and included sexual remarks, threats, false statements, name calling, and 'upsetting messages'. Given the ubiquity of cell phones in South Africa, this finding is disturbing but perhaps unsurprising.

What was perhaps a more disturbing finding in the UNISA study, however, was that over 40% of the respondents indicated that they were 'against reporting bullying', either because they were afraid of getting into trouble, or because they were afraid of further victimisation.

Siewierski, C. 2013

Although the second version of the text is still in relatively long paragraphs, with some unfamiliar words, you should have found it easier to follow. The only difference between the first and second versions is the use of paragraphs, but you might agree that it is easier to read and understand.

 When you read through the extract, were you able to understand all of the words and phrases used? The words 'cyberbullying', 'anecdotal', 'superficial', 'ubiquity' and 'victimisation' may be unfamiliar. Look up these words, or others that you are unsure of, on Oxford's online dictionary for adult learners (available at https://www.oxfordlearnersdictionaries.com/), and ensure that you understand them in the context that they appear in the extract.

Okay, you should now be pretty clear how important it is to use paragraphs to help organise the expression of your ideas and to aid reading. But when you write paragraphs, are you:

• comfortable deciding how to organise your information into paragraphs?

- able to clearly identify the topic and supporting sentences in your paragraphs?
- aware of what you are trying to do in each paragraph?

If you answered 'no' to any of these questions, it might be useful to note that paragraphs can be used for a number of different purposes. They can, for example be used to support, to concur, to qualify, to state, to negate, to analyse, to describe, to contrast, to synthesise, to summarise, or to expand ideas – but they can also be used to transition to a different idea. If you know what you need each paragraph to accomplish, it will be much easier to identify your main and supporting points, and Sections 7.2.1 and 7.2.2 will help you develop the first two skills.

 For lecturers, there is more on topic sentences and transitions available on the Learning Zone.

7.2.1 First things first ... paragraph architecture

Before you start writing paraphrases and summaries, it is essential that you understand the general 'architecture' of a paragraph.

A **paragraph** is a group of sentences that focuses on one main idea.

A paragraph is one main idea made up of smaller clusters of individual ideas (sentences). If this sounds confusing, let's look at Figure 7.1 below, which is a visual breakdown of the 'architecture' of a sample paragraph consisting of three sentences.

> Remember: A **sentence** is a group of words that together express one complete thought. A **sentence** must include a subject (a noun or proper noun) and a predicate, which includes at least a verb (showing an action or state of being) but often includes other elements as well. For example, 'Waiting for the rain' is not a sentence because it has no subject: we don't know who is waiting for the rain. They are waiting for the rain' is a complete sentence because it has a subject ('they') and a predicate ('are waiting for the rain').

Figure 7.1. Architecture of a paragraph

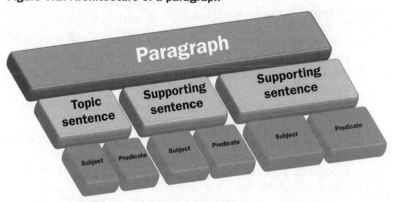

As you can see from the above figure, a paragraph consists of a topic sentence and supporting sentences, all of which relate to the main idea of the paragraph. These sentences may vary in importance and length but, in academic writing, they should each include at least a subject and a predicate. Because sentence lengths vary, paragraphs themselves can either be short or long, depending on the topic and purpose of your written piece.

> For more information on types and structures of sentences, your lecturer can give you the extended writing activity on the Learning Zone.

As you progress up the academic ladder, you are likely to find that your paragraphs get more complex and longer, as in the example we used earlier (with or without paragraphs). In general, try to keep your paragraphs short, and don't allow them to extend over more than 10 lines.

A structurally sound paragraph usually includes three components, as the example in the textbox below shows:

- **A topic sentence**
  > The main idea of the paragraph – conveys the writer's argument;
  > Usually the first part of a paragraph (but can be positioned elsewhere for effect);
  > Includes two parts: a topic + a controlling idea:
    - The topic (t) is the subject of the paragraph.
    - The controlling idea (c) makes a specific comment about the topic and limits the topic to a particular aspect that can be completely discussed in the space of a single paragraph.
  > Guides the writer on what to include, and the reader on what to expect;
  > Always a complete sentence (i.e. has a subject and a predicate);
  > *For example:* Completing pedagogically sound online activities (t) helps to improve students' comprehension of theoretical content (c).
- **Supporting sentences**
  > Develop the main idea;
  > Usually come after the topic sentence;
  > Make up the body of the paragraph;
  > Provide examples, detail, facts, evidence, statistics or explanation;
  > The order of supporting sentences can be changed in order to make most sense;
  > Each supporting sentence must be clearly related to the topic sentence it is trying to support;
  > *For example:* By completing these activities, students translate academic comprehension of theory into practical, concrete skills that are of direct value to learning and studying, and are also able to access prompt feedback on this understanding (Conrad and Donaldson, 2012). Additionally, other research (Conole et al., 2004) shows that use of appropriately designed online activities may also help students to identify areas of difficulty that they may not have identified in their reading of the theoretical components.
- **Concluding sentences**
  > Ending that either summarises the main points or rephrases the topic sentence, adding in a signal or clue that the

paragraph topic is now complete, and the writer is moving on to a new idea;

> *For example:* The activities can therefore reduce difficulty through further practice, and assist students in their creation of meaning.

Can you see that the word 'therefore' signals that this particular point is 'proven' and therefore complete, and that the writer is going to move on to a different idea?

You should now have a good understanding of the elements that make up a paragraph. But what makes a paragraph 'good' or 'bad' might still be missing from your understanding. Simply put, a good paragraph has both **unity** and **coherence**. By unity, we mean that each paragraph should discuss just one main idea, while coherence means that your paragraph should have a high level of readability and flow. To achieve this flow, you need to put your sentences in a logical order, and ensure that they are appropriately connected.

7.2.2   Connecting ideas in and between paragraphs

To ensure that your paragraphs and ideas have coherence, it is important to order your sentences logically, and to use suitable connectors and transitions between your sentences.

Before we explain how to do this, try to think back to a time that you revised an essay in detail. If you have done this, you might be familiar with the torment of a sentence just 'not fitting', and the subsequent time-eating torture of cutting it from one place, pasting it to another, reading, feeling dissatisfied, then cutting and pasting again, until you find the most 'comfortable' spot for it. Sometimes, this infuriating revision simply results in the sentence back where it started!

You might *also* have seen one or two of the following comments from your teachers or lecturers on your earlier assignments: 'choppy', 'flow', 'abrupt', '???', 'confusing', 'train of thought!'

If you are familiar with either of the above two scenarios, you are not alone. Many writers struggle to find the right 'fit' and flow in their writing – especially those writers who really value delivering good quality writing.

Getting your paragraphs to fit and flow in a way that feels right can be difficult, but it remains a really important part of the writing process. This section will present some of the most common approaches to sequencing your sentences (and paragraphs) and ensuring that they flow logically.

In the early stages of your academic writing career, it may be safer to always start your paragraphs with topic sentences, so that you can focus on sequencing your supporting and concluding sentences, and joining them using appropriate transitions. Table 7.1 below outlines some of

Transitions are words and phrases that help to establish logical connections between sentences, and between paragraphs or sections of writing. They provide signposts to readers about what they should do with the upcoming information. For example, transitions like 'similarly' or 'in addition to' alert the reader to the fact that a further example or statistic is coming up that will support what has already been said, while transitions like 'despite this ...' tell the reader that there is additional evidence or another point of view to consider.

the common sentence arrangements and their associated transitions (which can also be applied to the way you order your paragraphs!).

Table 7.1. Common sentence arrangements and transitions

Arrangement	Use	Common transitions
Chronological order	*Arrange supporting sentences in chronological order (from furthest away in time to most recent) or reverse chronological order (from most recent to most historical).*	Firstly, ... Lastly, ... At the start of ... More recently though, ... As a result of ... Because of ... Finally, ... Currently, ... Subsequently, ...
Order of importance	*Arrange supporting sentences in order from most important to least important, or from least important to most important.*	More importantly, ... Less significantly, However, ... Importantly, ... Significantly, ... Less notable ... Indeed, ...
Causation	*Arrange supporting sentences with cause followed by effect, or vice versa.*	Due to ... Because of ... As a result of ... So, ... Therefore ...
Comparison or contrast	*Present an argument for something then balance it with an argument against the same thing, or vice versa. Supporting sentences should be given for each part of the argument (for and against).*	However, ... Despite ... In contrast, ... Although ... In spite of ... Notwithstanding ... On the one hand, ... On the other hand, ...
General to specific or vice versa	*Present a general statement followed by detailed sentences, or present a detail and follow it with a more general concluding sentence.*	To a greater degree, ... On closer inspection, ... For example, ... On the whole, ... Thus ... Accordingly, ... To illustrate ...
Equal importance/ Examples	*For statements that all equally support the topic sentence, order is unimportant – the connections between the sentences need to show their equality though.*	And ... In addition to ... Similarly, ... Over and above ... Also ... In the same way, ... Equally important ... Furthermore, ... Just as ... so too ...

Refer to this table when you are writing essays, since it can provide you with practical ideas for organising your paragraphs and sentences. Just remember that there are many other ways to arrange your ideas and transition your paragraphs – the ones represented here are just some of the most commonly used ones. Considering each paragraph's purpose by examining the topic and supporting sentences should guide you in the direction of a suitable approach.

 Check out the student activity on the Learning Zone (LZ 7.1), the online presentation (LZ 7.2) and the video and activity (LZ 7.3).

7.3 Paraphrasing to represent another author's ideas

To **paraphrase** something means to retell or reword a piece of oral or written communication, often in order to make it more understandable to a different audience. If this definition is too abstract, think of it this way: you paraphrase every day, probably in several ways. For example, when you tell a friend what happened in a movie you watched, or in a fight that you saw, or when you tell your parents what a lecturer said in class, you are paraphrasing. In paraphrasing, you convey what the sources said, but in a different way. The word paraphrase comes from a Greek word that translates as 'to tell with modification' (Oxford University Press, 2019).

We paraphrase oral and written communication to make it simpler but there are other reasons to use it too. You may recall from Chapter 2 that it is really important for students to avoid plagiarism by referencing all of their work. An important part of avoiding plagiarism is to use your own words to retell another author's argument, i.e. to paraphrase your source. Copying an author's argument is *still* plagiarism even if you have indicated the source, so you need to reword and rephrase the original, while making sure that you retain the meaning of the original.

Direct quotations obviously need to be repeated (in quotation marks) exactly as the author published them, as well as being appropriately cited, but do not overuse them. Overusing quotations will not impress your lecturer. In fact, most educational institutions have clear policies that only allow for 5 or 10% of your paper to be direct quotes. Repeating what someone else has already said adds nothing to the academic world, and may give your lecturer the idea that you have misunderstood the task or the sources.

Instead, it is the way that you integrate and interrogate ideas from different authors, using these ideas to support your own argument, that is important. Paraphrasing also allows you to structure your own writing in a more logical way, since you can adapt your sources' ideas into your chosen structure (refer back to Section 7.2.2 for examples). The result is that your ability to paraphrase may be an essential skill to learn in academia.

7.3.1 A method for paraphrasing

The five-step method, presented in Figure 7.2 below, may be useful to get you started on the technique of paraphrasing, if you are not yet familiar or confident with it.

Figure 7.2. A five-step approach to paraphrasing

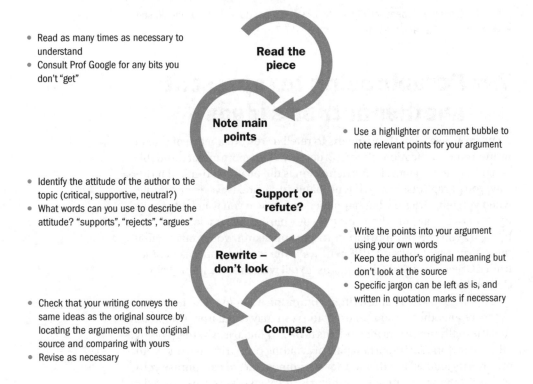

- Read as many times as necessary to understand
- Consult Prof Google for any bits you don't "get"

Read the piece

Note main points

- Use a highlighter or comment bubble to note relevant points for your argument

- Identify the attitude of the author to the topic (critical, supportive, neutral?)
- What words can you use to describe the attitude? "supports", "rejects", "argues"

Support or refute?

Rewrite – don't look

- Write the points into your argument using your own words
- Keep the author's original meaning but don't look at the source
- Specific jargon can be left as is, and written in quotation marks if necessary

- Check that your writing conveys the same ideas as the original source by locating the arguments on the original source and comparing with yours
- Revise as necessary

Compare

Find out more about the kind of reading suitable for preparing to paraphrase. See student exercise LZ 7.6 on the Learning Zone.

As you become accustomed to reading and using authors' ideas in your own writing, you will integrate the steps in the five-step process, and may stop noticing distinct phases. For now, try to complete each stage as comprehensively as possible, before moving on to the next phase, so that this practice slowly becomes automatic.

7.3.2 Tips for paraphrasing

This next section outlines some useful tips for paraphrasing that you should also consider when you follow your five-step procedure. It is important to note that none of these tips, on its own, is going to help you paraphrase content successfully. However, using them together and appropriately may help to improve your paraphrasing skills immensely. Again, this requires practise, so don't forget to visit the Learning Zone link for this section when you have covered the work itself.

Using synonyms

Using synonyms is an important part of paraphrasing, but note that simply replacing some words in a sentence with similar words does not mean that you have paraphrased the sentence. Paraphrasing is a broader concept, requiring a renegotiation of meaning; it often results in the comprehensive restructuring and reordering of an original piece of writing. But you can still use synonyms, or synonymous phrases, to convey the author's original idea in a different way (remember that you will still need to cite your source!).

For example, instead of saying 'Smith (2013) <u>rejects</u> the ...', you could say 'Smith (2013) <u>dismisses</u> the...'. But don't get carried away – using a synonym for a word in the original is not enough to be considered paraphrasing!

Changing word forms

Paraphrasing may also require that you change the parts of speech in a piece of writing. For example, you may need to change a verb to a noun, or change the tense or format to suit your sentence. If this sounds a bit confusing, realise that you probably do this all the time in your everyday life, without noticing.

For example, if we look at a sentence from Miller (2007:195):

> *'Surveys by the World Resources Institute (WRI) indicate that over the past 8 000 years human activities have reduced the earth's original forest cover by as much as 50% ...'*

We might wish to change this around, and instead say:

> *Indications from World Resources Institute (WRI) surveys show that a possible 50% of the earth's original forest cover has been lost due to human activities (Miller, 2007).*

In the original sentence, the word 'indicate' is used as a verb in the predicate, and 'surveys by the WRI' is the subject. In the paraphrased version, we have converted the verb 'indicate' into a noun ('indications'), which is now the subject of the sentence.

Point of interest

Changing a verb into a noun like this is called **nominalisation**, and is one of the ways in which we can make our writing more formal or academic.

Reordering main ideas

Unless you have been directly instructed to write a précis (a type of summary that follows the original order of the information), you are free to reorder information and arguments to suit your essay, as long as the new order does not prejudice or change the source's intended meaning in any way.

Switching to active or passive voice

You may remember the active and passive voice, mentioned in Chapter 4. The passive voice should be reserved for more formal pieces of writing because it tends to complicate and lengthen any sentence. You may find switching between these voices useful for paraphrasing, especially if it suits your chosen style of writing.

For example, in an original passive voice sentence, Miller (2007:105) says:

> 'Heat is absorbed and released more slowly by water than by land.'

In this sentence, you may have noticed the word 'by' signalling the passive voice, and seen that the first noun in the sentence ('Heat') is actually not the subject of it. 'Heat' is having the action performed on it, rather than performing the action itself. If we reworked the sentence to use the active voice, it could read:

> Miller (2007) argues that water absorbs and releases heat more slowly than land does.
>
> OR
>
> Miller (2007) notes that land absorbs and releases heat more quickly than water does.

Can you see that we have used exactly the same information, but converted it to active voice in two different ways? Because we have understood the full meaning and implication of Miller's original statement, we have been able to manipulate the information in different ways. By changing the original from passive into active voice, we may also have simplified the author's original statement.

Let's bring these tips together – a practical example

To bring all of these tips together in practice, let us examine an earlier example by Siewierski (2013):

> What was perhaps a more disturbing finding in the UNISA study, however, was that over 40% of the respondents indicated that they were 'against reporting bullying', either because they were afraid of getting into trouble, or because they were afraid of further victimisation.

If we apply all of the above tips to the original sentence, we could come up with something like this:

> Siewierski (2013) notes, however, a finding of even greater concern. The author (Ibid.) cites the findings of a UNISA study, in which an excess of 40% of sampled respondents rejected the practice

of reporting bullying incidents. While some cited a fear of reprisals, others cited a fear of further persecution.

Although we did not change the voice in this example, can you see how the sentences have been completely reworked, and rearranged, and that we have used a number of different synonyms and synonymous phrases? You will also note that we brought the original author directly into the sentence to show that we were citing a source!

This is one possible paraphrase of the extract – but there are literally hundreds of ways in which it could be rephrased. It is up to you to adapt the process and the tips so that they work for your style of writing. If you remember to cite your source, and use your source's arguments in your own voice to *support* your argument, then you should be able to keep a clear idea of how and when to paraphrase.

Practise your paraphrasing skills and show your understanding of why we paraphrase with student activities LZ 7.4 and 7.8 on the Learning Zone.

7.4 Writing summaries

The ability to write a **summary** (an abbreviated or shortened version of an original text) is another essential tool in your academic skills-kit. For example, think about how frequently you need to summarise ideas: you take summary notes of discussions in class; you make summary notes for studying; you make summaries when you are trying to find information for an assignment; and you may even be required to write specific types of summaries for different classes, depending on which courses you take.

This section will outline some of the most notable features and types of summaries, and provide you with a solid approach to making your own summary notes.

7.4.1 Features of a summary

Summaries differ depending on their uses, but many summaries share similar features. For example, a summary:

- will vary in length depending on its purpose and the length of the original document – if no specific length for the summary is given, it is generally 'safe' to aim for a third of the original;
- focuses on accurately reducing the main points of the original text;
- leaves out supporting details and examples;
- requires the writer to paraphrase – it does not use the original author's words verbatim (or as direct quotations);
- may change the focus of the original document, depending on its use;
- may change the order in which information is presented;

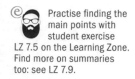

Practise finding the main points with student exercise LZ 7.5 on the Learning Zone. Find more on summaries too: see LZ 7.9.

- considers the reader's needs and formats language and sentence structure accordingly;
- may change the format of the document, or use mind maps or other summary diagrams.

If you think about the notes that you write for yourself in a class, interacting with a text, or studying a subject for an exam, you may realise that you make use of your own shorthand, such as abbreviations or symbols, instead of writing out commonly used words. This kind of summary may be useful to you, but it would be useless to a group of your classmates if you all decided to contribute a summary of one chapter of a textbook. Similarly, if you needed to provide your lecturer with a short summary of a project that you are engaged in, your language and approach would be somewhat different.

Remember then that there are different types of summaries, used for different purposes. If you use your own words, consider your audience, and follow any specific instructions given, your summary should be acceptable.

7.4.2 Précis and abstracts

Précis

You may remember us referring to précis in the earlier section on paraphrases, or recall having to write them at school for English. A **précis** is simply a summary, usually about a third of the length of the original, but it differs from an ordinary summary in two ways: because it leaves the information in the original order that it appeared, and because it maintains the proportions of the original.

If you are not studying a language at university, you are unlikely to be asked to write a précis, since ordinary summaries are more frequently used.

Abstracts

An **abstract** is a short summary of a study and its findings that appears on the first page of journal articles, research papers, dissertations or theses. The informative abstract (a more common and complete type of abstract) is used to outline the main points and findings of a study and can be lengthy, in contrast to the less common descriptive abstract, which provides just one or two short sentences.

An informative abstract allows potential readers (like you) to quickly determine whether or not the study will be worthwhile or relevant enough to read. Reading an abstract first is worthwhile because it can save both time and money – many journal articles require payment before access. In your early years of study you may not have to write an abstract of your own, but you will need to read and understand them in order to source relevant content. Familiarising yourself with them will also be helpful if you further your studies later.

7.4.3 Summary-writing technique

When you sit down to write a summary, you need to ask yourself a number of questions before you start typing, or put pen to paper. If you incorporate this into a set sequence, you can try to ensure that your summary will suit your purpose, and record the original document in a relevant and reliable way. We therefore recommend that you follow this eight-step procedure:

1. **Identify your reader**
   - Know who you are writing for, what kind of language they are expecting, and what length of summary is suitable;
   - If it's for yourself, you can use your personal shorthand and symbols and so reduce the length of the summary.

2. **Identify your purpose**
   - If you are writing a summary of a presentation for your classmates, it will not work to use your own shorthand;
   - Look at the length of what you need to get through and the level at which you need to be able to discuss the topic – summarising a chapter is likely to yield more detail than if you have to summarise a whole book for study purposes, for example.

3. **Read the original twice to find the main ideas and highlight them**
   - As with paraphrasing, it is vital that you understand what you are summarising;
   - Highlight the important headings, subheadings, topic sentences and important points with actual highlighters, or by using electronic comment bubbles or highlighters, depending on where you are looking at your information;
   - Use Professor Google or other textbooks to help clarify points you are struggling with (but try to avoid getting sucked into irrelevant but fun Internet posts!).

4. **Write down the main ideas in your own words**
   - Writing the main points in your own words helps to clarify anything you may have been confused by, and also forces your brain to process the information, as opposed to simply trying to recall it.

5. **Decide on a logical order for your main points**
   - Decide on a suitable structure, such as a mind map, or an essay with headings and subheadings, depending on what type of information you are dealing with, and what your reader is expecting.

6. **Write a topic outline or draw a mind map, depending on what you need**
   - Use headings, subheadings and topic sentences to order your plan according to suitable levels.

7. **Check your outline against the original and your notes**
   - Ensure that you have not missed anything important;
   - Check that your ideas are in a logical order;
   - Make sure that your summary is a suitable length, and that it has the right level of detail (not too much, and not too little) for your purpose.

8. **Write the final version**
   - If the summary is for your use, simply write it out as neatly as possible in a way that will work for your needs;
   - If the summary is for someone else, make sure that it uses complete sentences, that it uses appropriate, grammatically correct language, and that it delivers what it needs to. Run it through a word processor for spelling and grammar errors, or proofread it if it is a handwritten version;
   - Your final version should be an accurate representation of the main points, presented in a logical order, and at a length and format suitable for purpose and audience!

Try to use this process if you are not yet comfortable writing different summaries, even if it seems excessive. Many students struggle to use their own words in summaries; writing summaries for study purposes will give you great practice at doing this. Use these skills as often as you can with your prescribed course content, and try to improve your paraphrasing and summarising skills throughout your studies.

 Check the Learning Zone for more activities, e.g. how to tell the difference between a paraphrase and a summary (LZ 7.7), writing summaries (LZ 7.10 and 7.11), fun summaries (LZ 7.12).

7.5 Conclusion

This chapter focused on developing three key academic writing skills, the first of which was understanding the role of, and how to write, structurally sound paragraphs. As part of this, we identified common methods of sequencing sentences and using transitions, and used these in practical exercises. The chapter also provided reasons for using paraphrases and summaries, and offered practical approaches to these two forms of writing, both of which are vital for study purposes.

References

Conole, G., Dyke, M., Oliver, M. and Seale, J. 2004. 'Mapping pedagogy and tools for effective learning design'. *Computers and Education*, 43(1–2): 17–33. Retrieved 22/11/2019, available at <https://doi.org/10.1016/j.compedu.2003.12.018>

Conrad, R-M., and Donaldson, J. A. 2012. *Continuing to engage the online learner: More activities and resources for creative instruction*. San Francisco, CA: Jossey-Bass.

Fielding, M. and Du Plooy-Cilliers. 2014. *Effective business communication in organisations: Preparing messages that communicate.* 4th ed. Cape Town: Juta.

Miller, G.T. 2007. *Living in the environment: Principles, connections and solutions.* 15th ed. Boston, MA: Cengage Learning.

Oxford University Press. 2019. *Oxford learner's dictionaries.* Retrieved 30/08/2019, available at <https://www.oxfordlearnersdictionaries.com/>

Siewierski, C. 2013. *Managing cyberbullying in South African schools: An analysis of St Mary's anti-bullying policy.* Unpublished paper for Leadership and Management of Teaching and Learning Module, University of the Witwatersrand.

Thoreau, H. D. 1854. *Walden.* Boston, MA: Ticknor and Fields.

CHAPTER 8

Critical analysis and argumentation

'If you would be a real seeker after truth, it is necessary that at least once in your life you doubt, as far as possible, all things'
– René Descartes (2009:130)

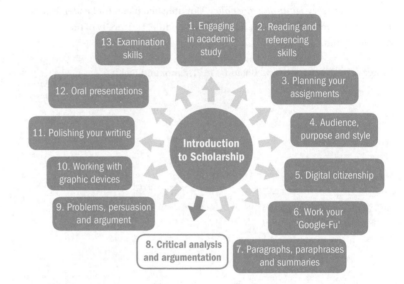

13. Examination skills

1. Engaging in academic study

2. Reading and referencing skills

12. Oral presentations

3. Planning your assignments

11. Polishing your writing

Introduction to Scholarship

4. Audience, purpose and style

10. Working with graphic devices

5. Digital citizenship

9. Problems, persuasion and argument

6. Work your 'Google-Fu'

8. Critical analysis and argumentation

7. Paragraphs, paraphrases and summaries

OBJECTIVES

At the end of this chapter, you should be able to:

· understand the subjectivity in the notion of 'truth';
· acknowledge the role that cognitive dissonance plays in our evaluation of information;
· recognise the importance in academia of adopting the lens of a 'critical observer';
· use a given method to critically evaluate evidence;
· detect cognitive bias and logical fallacies in arguments.

If a pleasant but odd acquaintance of yours casually informed you that everything you believe is false, and that what you know and experience as 'reality' and 'truth' are simply illusions designed to keep you enslaved to the Flying Spaghetti Monster, would you take their word for it?

What about if a lecturer you respected stood up in front of your class, and told you the same thing in a slightly more academic and formal way?

Would your lecturer's stature and history of credible research carry enough weight for you to abandon your old beliefs and adopt this new idea, or encourage you at least to reconsider your current beliefs?

And what if that same respected lecturer told you not to worry if you don't 'get' algebra – because you are obviously a 'right-brained' learner?

8.1 Introduction

In the first *IRL* example above, there is little doubt that you would quickly dismiss your acquaintance's claims about the Flying Spaghetti Monster. When a lecturer you respect offers you the same information, you might *entertain* the claim a little longer but there is little chance you will rewrite your entire conception of the world to fit this new (crazy) claim. You might, however, accept the third example. If you have read any articles or books on training different sides of your brain, the statement might have some credit; your lecturer, a person you admire, presented the idea, and whatever they say *must* be right, so it carries more weight; and finally, you might have struggled with algebra since school, and so you welcome the 'neat' solution your lecturer provides. Through no fault of your own, you feel, you simply cannot do algebra.

The first problem with your lecturer's statement is that it assumes that you are 'right-brained'. Secondly, it assumes that being 'right-brained' results in an inability to do algebra. Neuroscience – the science that studies the brain's functioning most closely – does not actually support a 'left-brain: accountant, right-brain: artist' view (Nielsen et al., 2013; Lilienfeld et al., 2010), and so the statement *should* be rejected.

So why do we dismiss some arguments immediately, entertain others and *then* dismiss them, or accept some ideas easily even when they are wrong?

The short answer is that the human brain uses a complex set of considerations to decide if what we are hearing is truthful and valid, complete nonsense, or something in between. When we analyse arguments then, we are using a combination of logic, emotion, and preconceived ideas and constructs to draw a conclusion that we feel is the most likely. Our so-called 'analysis' of an argument is therefore often shaped by:

- what we know (or think we know) about *who* is trying to convince us;
- what we (think we) already know about the *topic* from our personal experience, research, or casual exposure;

If you are curious about the Flying Spaghetti Monster (which is a real organisation), you can check it out at http://www.venganza.org/about/ *Warning*: The site is deeply satirical, and if you are religious, you might be offended by it, so visit with caution.

If you would like a little more insight into the 'left-brain right-brain' myth, visit this short but great YouTube video by Professor Dave Explains here: https://www.youtube.com/watch?v=mKlkwJypzJQ

- how strong our current beliefs on the topic are;
- and by how 'reasonable' the argument *sounds* (which is subjectively determined by a number of things, including our world views, our preconceived ideas, our degree of open-mindedness, and the structure and approach of the argument itself, including its use of rhetoric.)

Don't worry about **rhetoric** in this chapter other than to note that it is about persuasive language – we will deal with it in detail in Chapter 9 under Persuasion.

If you noticed that considering the quality of the statement itself and its supporting evidence doesn't feature in the list of things shaping the analysis, give yourself a pat on the back! Logically, these would be the most essential elements to consider. But because of fixed thinking, logical fallacies, the successful use of rhetoric, and the human tendency to allow emotions to affect how we evaluate arguments, the very thing that we *should* be evaluating – the quality of the statement itself – is often under-analysed, or even ignored.

For example, if we had been alive in 330 BCE when Aristotle theorised that the Earth was a sphere, we would likely have rejected this 'crazy' idea outright because we already 'knew' that the Earth was flat. Even though Aristotle presented solid evidence, and Pythagoras had come to the same conclusion a few hundred years earlier, we might have rejected the idea because it conflicted with our existing 'knowledge' of the world.

This kind of 'messy', incomplete, and often stubborn or illogical approach to observations and argument will not do in an academic context, since the goal in academia is to establish and build reliable bodies of evidence. We therefore need to be objectively critical, and reflect honestly and consciously on the ways in which we evaluate evidence.

In the academic context, being **critical** means reflecting on both the negative and the positive elements of a statement – it does not simply involve finding fault with something. For example, 'Anna was asked to critically analyse Baz Luhrmann's film adaptation of Romeo and Juliet for her English Literature assignment. She noted that it was a rather dramatic adaptation, but acknowledged that Luhrmann had been faithful to the themes of the Shakespeare play itself.'

This chapter will outline the importance of adopting a critical, conscious lens in academia, will look briefly at some of the basic concepts around 'truth' and how we find it, and will then guide you through a method of critical analysis to ensure that you consider other people's arguments fully and critically. Before we don this 'critical lens' though, it is important that we first deal with the concepts related to 'truth' and critical thinking that underlie this approach, and explain why humans, in general, are resistant to changing their minds – sometimes even in the face of solid evidence. By identifying different types of statements and the roles that emotion, context, rhetoric, and logical fallacies play in our analysis of arguments, this chapter will help you to evaluate the arguments of others so that you can make reasonable decisions about what to believe, and which courses of action to follow. Importantly, this chapter will also lay the foundation for Chapter 9, which will lead you through writing your own 'bulletproof' academic arguments.

8.2 So just what is the 'truth' anyway?

In the earlier example of Aristotle trying to convince the world that the Earth was spherical, we pointed out that we would be likely to have rejected his world-shaking idea. Similarly, if a group of people now tried to convince us that the Earth is flat and not spherical, we might laugh and dismiss their claims – we know that it is spherical!

But *do* we know? And if so, *how* do we know? Can we actually *feel* the Earth rotate? Have you personally *observed* that our Earth is spherical? Or do we know it just because science-y people, supposedly smarter than us, have told us so during our education?

Although you need not worry about this particular (flat-Earth) claim, the questions above *should* raise some interesting points about *what* we know, and *how* we come to know and accept ideas. How do we make sure that what we accept as fact or knowledge is actually true? And how do we know what the 'truth' or 'reality' is anyway?

 Test your ability to differentiate between fact and opinion by doing an online assessment (Chapter 8 Fact or opinion) on the Learning Zone.

8.2.1 Life through a lens

A first step closer to 'reality' is to do away with our **naïve realism** – the (mistaken) assumption that we see the world exactly as it is (Ross and Ward, 1996). Our 'common sense' might tell us that we have eyes and ears, so surely, we should be able to see what goes on around us and accept what we experience as fact? Shouldn't we?

Well ... yes, it *is* true that we experience the world through our senses, but actually, the unconscious 'sensing' part of it is really just the first step. Assuming our sensory organs have done their job and collected sensations, it is actually our brain (our very own supercomputer) that does the job of deciding just 'what' it is that we have seen, heard, touched, smelled, or tasted. It is a fascinating process; while we cannot go into detail here, our brain's job, in short, is to take the flood of signals sent from our sensory organs, select which of these signals to prioritise and which to ignore, decide where to store and organise the information, and then, finally, decide on how to interpret and act – if necessary – on the basis of these signals.

Although that process takes just milliseconds, it is complex, since it involves a number of steps, signals and possibilities. Importantly, as McLeod (2010) argues, the results of sensory perception are deeply influenced by a number of factors, for example: whether we are processing something consciously or unconsciously, what mental models (or learning, attitudes, perceptual biases or expectations) we have about its elements, and even what our values, current mood, and motives are. These factors explain, for example, why eyewitnesses at crime scenes often report seeing wildly different things, even if they did see the same crime and believe they are reporting it accurately and honestly. It also explains why the same plan might look brilliant and logical to some, but completely ridiculous to others.

How do we get around this to arrive at some sort of 'agreed reality'? In academia, one of the ways we avoid naïve realism is firstly to understand and acknowledge how our worldview, academic focus, and interests might affect the data or information we are observing or generating. We can refer to the things that influence our perception of

Point of interest

Despite what flat-Earthers might claim, there is abundant evidence that the Earth is a sphere – among other things: GPS devices, photographs and videos of Earth from various satellites. And, before doubters shout 'Adobe Photoshop!' or 'GPS rigging', you can also observe the curvature of the Earth yourself by looking at the Earth's shadow during a lunar eclipse, watching a ship disappear on the horizon, or taking a flight around the world.

For more information on **mental models**, visit Sections 9.2.1 and 9.2.2 in this book.

reality as our 'lenses'. To understand how this works, Makhmaltchi et al. (2017) suggest imagining that that your mind is a tall, solid wall full of lenses – and behind this wall, lies Reality. You cannot see over the wall, so your only way to see reality is *through* those lenses. Each lens, like different prescription glasses, telescopes, microscopes, or various shades of sunglasses, focuses on a different perspective or element of the truth beyond.

Let's use an example to demonstrate. We will keep with our earlier crime-scene theme and assume that the reality beyond the wall includes an armed robber and three (bloodied) victims. If we look at reality through one of our lenses, we see the impact of what the robber has done to the immediate victims. Through another lens, we see the impact of the crime on greater society. Another lens shows us what factors led the robber to this life of crime; one further down looks at the economy's role in crime; and one higher up reveals how the legal justice system is skewed to those who have money to defend themselves. All of this from just one 'reality', and we haven't even looked through all our available lenses yet! So which lens is right?

Unfortunately, none of our lenses offers a complete, reliable depiction of reality. Each of them suffers some distortion, focus or bias, because our academic disciplines, personal interests, world views, values and beliefs all play a role in what lenses we use and how we use them. To demonstrate the role of values, if you think justice should be about balance and fairness, you might normally look at the robbery and hope that the justice system acts quickly by restoring safety and ensuring that the robber is humanely punished and rehabilitated. If you or a loved one has been a victim of violent crime, you might feel angry, and hope to see swift and severe punishment of the robber to remove them from society and discourage others. Or, if you were someone else, you might take a more radical view that the criminal justice system itself is broken because a few wealthy people have all the power and means, and the powerless have few options but to resort to criminality. Your hope might then be to dismantle and correct the system, rather than to see punishment or rehabilitation of the robber.

Aaaaand – we're not quite done yet. Finally, in academia, we would *also* need to acknowledge the theoretical or disciplinary lenses we might be using. As is demonstrated in some of the lenses to look at the robbery, you will encounter and use different lenses depending on the competing theories that inform your discipline, because sociology, criminology, biology, science, or physics, for example, each have their own lenses, which at times compete, at others complement one another. As you move up the years of study, you will find these lenses become increasingly important.

So, if there are so many lenses, and each person has different lenses that we don't even share, is there any hope at all for us to 'find' reality? Is there any point?

Absolutely. It is impossible for us to examine every element of every-thing in a purely factual way – that is simply not the nature of social creatures or their contexts. We and our world are complicated, tangled webs of information and influence. To use a cliché, using an academic lens might be compared to eating an impossibly big elephant, one bite at a time. Academic lenses help us to manage unimaginably big data sets. Being aware of and acknowledging the lenses through which we view the world, and communicating our thinking and findings clearly and honestly, can help us to identify and frame problems and possible solutions from different perspectives. Perhaps most importantly, as Makhmaltchi et al. (2017) might agree, the existence of different lenses and our willingness to explore them can help us to 'stitch' together as complete a picture of reality and the external world (in other words, the elephant!) as possible, so please do not give up your quest for The Truth just yet, oh Fearless Student!

8.2.2 Ontology and epistemology

As you will have gathered from the above section, the concept of 'truth' or reality is, in itself, a tricky notion to explain. There is no universal agreement about just what the 'truth' is, so in academia, you will find different definitions depending on which school of thought or paradigm you follow. Ontology and epistemology are concepts that you will encounter frequently in this regard. Um, what?

In this context, a **paradigm** refers to the framework of theories, concepts, research methods, assumptions, values and tenets of a particular field or school of thought.

Simply put, **ontology** refers to the truth*, or nature of reality* (* Terms and Conditions apply ...), while **epistemology** refers to how we 'know' the truth*, and what methods we use to *find* it. Here's the catch: different perspectives have wildly different ideas of just what reality *is*, and how to find it, so a single definition of ontology or epistemology will not really help. Examples will help though, so in this section, we will outline how three main academic perspectives see ontology and epistemology. Although there are other viewpoints, positivism, interpretivism, and critical realism are a good place to begin.

Positivist ontology argues that 'truth' is real, external, and can be observed and measured. Whether we believe it or see it, or not, it exists outside our influence or perception. In other words, the truth is the truth is the *objective* truth – and everyone should agree on it.

Most of us would accept, for example, that there is little to argue about in the sum $2 + 2 = 4$. We would also likely agree that gravity exists as a force, even if we did not understand the science behind it, simply because we observe and experience gravity's effects all the time: we don't float into space when we take a walk outside, balls that are thrown up in the air come down, and, when we stand on a scale, our mass offers us evidence (not always welcome!), that gravity does, in fact, exert a force on us.

For a positivist, it is the job of science to identify truth, measure it, and record it so that we can analyse it and make predictions for the future.

From the positivist view, data about the 'truth' can be collected in an objective way to find universal laws of cause and effect, to generalise findings about them, and to test and make predictions about the future. It is empirical. And, while it should not pretend to have all the answers (yet), the positivist approach does hold that reality is ultimately 'knowable', and that we simply need to find the right question, and then use the right tools and approach, to measure it. For these reasons, you will find the positivist approach is very popularly reflected in the natural science fields – and it is easy to see how their precise measurements have helped sciences like physics to advance.

Okay, so the positivist view of 'truth' sounds rather neat and tidy. Unfortunately, this conception of the 'truth' is rather more difficult to apply to social settings, perhaps because humans and social interactions are less inclined to follow universal laws of nature, and are more inclined to follow dynamic and complex communication models! For example, it would likely be impossible for science to accurately predict the effects of standing on a scale for the first time and discovering their mass on someone's self-esteem. Humans, and a lot of the world that they inhabit, are simply more complex than the application of cause and effect, or universal laws of nature.

For this reason (and a number of others), followers of the interpretivist paradigm reject the idea that the 'truth' is an external and objectively examinable thing. They argue instead that 'truth' is rich, complex, socially constructed through language and culture, and that one person's experience of reality will not be the same as another's. This includes researchers, so empiricism is a farce, and reality does not exist!

While we might agree with their criticism of positivist assumptions of objectivity, and can certainly value interpretivist efforts to describe, analyse, and interpret these 'relative truths', their rejection of reality as an idea does somewhat limit our ability to apply interpretivist research to the wider world.

Enter our third contestant, critical realism. Critical realists, like good (but complex and cheeky) diplomats, have recently entered the academic fray to say that positivists *and* interpretivists were right about truth and how we find it. How? Surely these positions are irreconcilable? Well, critical realists *also* argue that both approaches are more than just a little bit wrong too. The philosophy is more intricate than explained here, but critical realists (Bhaskar, 2008) argue that much of our reality exists and operates with or without our knowledge of it. Reality is comprised of layers (or domains) and structures: some we can see, and some we cannot. They agree that an external reality exists (score one for the positivists!), but they disagree that we can know or measure it all, or that what we measure is objective and accurately reflects reality (score one for the interpretivists!). Instead, critical realists argue that by acknowledging limitations and biases and working consciously to reduce them, researchers can aim to be objective and try to find as many layers of truth as possible through examining structures. Importantly, if we understand and describe the structures and relationships that affect reality, we can make positive changes to a rather unequal world, and give a voice to the voiceless too.

Okay, so where does this leave the 'truth'? We can use an example to demonstrate the difference between these perspectives. Let us assume that Country X has a large problem with corruption in government spending, but that we are not quite sure exactly how big the problem is or why it exists. The positivists could deal with one element of this problem swiftly. They may, for example, conduct an independent financial audit of all irregular expenditure across all government departments in Country X over a period of one full financial year, from 1 March 2020 to 28 February 2021. Through this systematic, measured and quantitative process, positivists would determine how much money has been lost to corruption, and provide us with a particular figure, or at least a reasonably close estimate. However, they could not tell us how, or why, this money was taken, so we would be no closer to understanding the nature or causes of the problem.

Enter the interpretivists. They might, for example, argue that individual motives behind the corruption are much more interesting than a number. So, instead of using quantitative methods like data analysis, they might use a qualitative method and conduct in-depth interviews with a small sample of the identified perpetrators to establish their perspectives, and look at *why* they allowed themselves to be corrupted. In this way, they would try to find the 'truth' according to the sampled government official, and develop a rich, informative analysis of the individual's motivations. They would not, however, claim these motives as valid for all officials involved in the corruption, because their research would focus on the sampled individuals.

Critical realists, on the other hand, might explore the social impact that corruption in government spending has on particular communities. They often use a range of research methods to examine a single problem. In this case, they might use a quantitative research method like analysing the financial statements and social spending of a specific municipality. They could use these results to inform their next step, which could be a qualitative research method such as interviewing a sample of convicted officials using a set questionnaire to determine why they allowed themselves to be corrupted, or interviewing a sample of residents to determine their views on service delivery. Once the various research steps are complete, they might propose practical steps the municipality could take to avoid future losses to corruption, urge them to deliver specific services, or offer suggestions on the role residents can play in holding government officials accountable.

As you can see from this example, these approaches see 'truth' very differently, and attempt to find it in rather different ways. Each approach also has advantages and disadvantages. So, in addition to the specific critical evaluation method outlined in Section 8.4 below, we recommend that your approach to evaluating what statements are 'true' and reliable in your sources is best guided by your specific academic context. Although this is not a hard and fast rule (check with your lecturer first), if you are in a Natural Sciences or Engineering Faculty, you are likely to be safe trying a (new) positivist's notion of 'truth'. However, if you are part of Education, Business, Humanities or Social Science faculties, an interpretivist or critical realist view might be more suitable instead. Think carefully about the logic and usefulness of different paradigms as you progress through your studies – don't be scared to explore them more deeply.

For a great introduction to different research paradigms, read the paper 'Research paradigms and meaning making: A primer' by Krauss, available at https://nsuworks.nova.edu/tqr/vol10/iss4/7/. If you are interested in exploring ontology and epistemology further, we can also recommend *Research matters* by Du Plooy-Cilliers, Davis and Bezuidenhout published by Juta in 2014.

8.3 Critical thinking and cognitive dissonance

Unless that previous section sent you running for the lesser demands of the Twittersphere, you should now have a better idea about the difficulties in pinning down the 'truth' or reality. This is important, because we now come to the business of adopting a critical lens, or approach, to evaluate claims. Simply put, critical thinking is a solid way to identify which claims are the most reliable so that we can adopt and develop them, and discard any weak or unreliable ones. It is important that a critical approach starts with Descartes' rather uncomfortable advice to '... doubt, as far as possible, all things' (2009).

Doubting 'all things' is difficult for most of us, because we tend to feel secure in the things that we think we know. However, to think critically, we need to take the risk of stepping outside of what we think we know about the world, and reject claims at face value. Instead, we need to *actively interrogate* an idea by asking *what, when, who, where, how* and *why* questions, and by recognising faulty logic, generalisations, ambiguity, or other errors in argumentation. It is only once we have done this that we can assess an idea's apparent merits, and arrive at a logical, rational conclusion about the idea's validity.

While this sounds logical enough, rational conclusions can cause turmoil and are sometimes difficult to accept. Critical thinking sometimes means needing to discard old ideas when they are proven wrong, or accepting strange new ideas that have more merit. Sometimes the results of critical thinking even require us to leave an uncomfortable theoretical 'gap' in what we thought we knew about something. This can happen when we realise that our evidence is incomplete, when current ideas simply haven't been tested yet, or when we do not yet have enough information to form a complete theory.

Festinger (in McLeod, 2018) argues that we humans tend to prefer our beliefs and opinions to be consistent with each other, i.e. we struggle with contradictory ideas, and are uncomfortable with information that suddenly contradicts what we already believe. So, when we find new information that differs from something we strongly believe, we experience something Festinger calls **cognitive dissonance** (Ibid.). This dissonance, or feeling of tension, occurs when a person holds two conflicting attitudes, beliefs or behaviours at the same time. For example, if you voted for a particular political party because you respect its policies, but recognise that the leader of the party is corrupt and bad for the country, you may feel quite uncomfortable when someone who supports a different political party points out the leader's shortcomings.

Festinger's theory (McLeod, 2018) says that we can get rid of this uncomfortable conflict or dissonance by opting to:

* change our minds and exchange an old idea for a new one (this is usually difficult, and so uncommon); or

---

Check that you don't confuse **critical thinking** with **critical theory**. Critical thinking refers to a particular approach to knowledge, while critical theory refers to a group of theories that see society as fundamentally divided by unequal power relations. For a more complete discussion on this, read Burbules and Berk's article 'Critical Thinking and Critical Pedagogy: Relations, Differences and Limits', available at http://mediaeducation.org.mt/wp-content/uploads/2013/05/Critical-Thinking-and-Critical-Pedagogy.pdf

Point of interest

The Theory of Everything (or ToE) in physics is a good example of this **'gap'** – it is a hypothetical theoretical framework that tries to tie together the theories of general relativity and quantum field theory, which are currently incompatible with each other. Our present understanding means that only one or the other of these theories can be right – not both – and yet both theories have experimental support. The ToE has therefore been created to try to find a possible scenario for how these two apparently irreconcilable theories can be united. As current research stands, 'string theory' looks like the most promising of these ToEs and may, in the near future, provide an explanation for how both general relativity and quantum field theory can be 'right'.

- find new information to try to support the old belief instead (which is a little like grasping at straws, but very common); or
- downplay the importance of the new idea (which is very common indeed).

Because we try to avoid further dissonance, Festinger (Ibid.) claims that we are more likely to try the middle and last options than the first one. This is especially true of ideas or beliefs that we value a great deal. For example, in our earlier example of the corrupt party leader, you could decide to agree with your opponent, and then vote for a different party, but you may be more likely to try to find examples of good deeds by the leader of your party; try to find examples of corrupt practices in your opponent's party; or tell yourself that the leader can only serve two terms as the leader, and so will be gone soon. You may also steer clear of thinking about the leader, avoid situations in which he might come up in conversation in order to avoid further dissonance, and continue to vote for the party. By doing these things, you will have shifted the dissonance, but not actually dealt with it.

Although it is sometimes difficult, as critical thinkers, we need to be aware of cognitive dissonance, and interrogate both why we feel uncomfortable, and what we need to do to fully resolve any conflict. Sometimes, we simply need to choose the tougher option – changing our minds. If we continue to take 'easier' options, making excuses, or reducing the importance of problems, we are much less likely to develop academically, or be able to differentiate between a valid argument and a specious argument.

Specious arguments are arguments that superficially appear right, but which are actually wrong.

Complete a guided reflection activity, using student exercise LZ 8.1 on the Learning Zone.

8.4 A step-by-step method for evaluating evidence

If you feel lost about how to decide whether an argument is valid or not, you might find it easiest to follow a process, especially at the start of your academic practice.

The good news is that you are already at least partially equipped to critically analyse different pieces of writing. This is because the SQ3R study reading method we described in Chapter 2 (Section 2.3.5), and which you have already mastered, provides an excellent platform from which to launch. You simply need to add in a few extra steps.

The **SQ3R method** is covered on pages 26–28 of this book. Make sure that you integrate this section's checks with this method to maximise your critical lens.

Firstly, when you are reading the content actively as part of *Step 3* in the SQ3R method, you will need to identify which of four types of statement you are working with:

1. **Analytical statements**: these are by definition true, and leave little space for arguing. For example, the example we offered earlier, that 2 + 2 = 4, is an analytical statement, because it can be proved. Obviously, this counts as strong evidence;

2. **Value judgements**: these reflect an opinion, and so depend on the perspective of the speaker. For example, a person may claim that Party X is much better than Party Y. Certain elements of this may or may not be true, but it is ultimately just an opinion and so does not qualify as 'evidence' – the speaker will need to provide a lot more evidence for this statement if they want you to agree with it;

3. **Metaphysical statements**: these are also a matter of belief, and, by definition, cannot be proven or tested. They offer speculation about things that we cannot observe, and so also include statements about belief. For example, if I claimed in an academic paper that the Flying Spaghetti Monster actually exists, you would be exercising sound academic judgement to discount my statement completely. Because we cannot measure metaphysical statements, they would only be appropriate in certain philosophical discussions or papers, and would then need to be supported by convincing logic or reasoning. These claims cannot, however, be taken as evidence in and of themselves;

4. **Empirical statements**: these cannot be proven, so instead, they offer theoretical likelihoods that rely on being denied. For example, if I said that there is strong evidence to support the theory of gravity, this would be an empirical statement. This is because we can use repeated tests and measurements to see whether gravity holds as a theory, although its existence cannot be proven in the same way that 2 + 2 = 4 can. Importantly, if just one single, reliable piece of evidence contradicts a significant amount of evidence supporting a theory, the statement of the theory would be found to be invalid or incomplete.

 When you are working through Step 5 of the SQ3R method, you will have a better idea of whether the author has provided fact, well-supported theory, opinion, or reasoned argument if you know which of the four types of statements they have used.

In addition to identifying the types of statements and completing the steps of the SQ3R method, using the checklist (in Table 8.1 below) throughout the process will add another layer of evaluation to your reading, particularly during Steps 2, 3 and 5.

 If you find that integrating the checklist (in Table 8.1 below) is too difficult at first, complete your SQ3R and then run through the arguments and claims that you identified, this time with the checklist. You will find that integrating these tools becomes more natural as you practise them, but do not be too concerned if you feel that you are analysing things to death and repeating yourself a lot in the beginning of your critical reading practice.

The good news is that if you have completed the SQ3R reading method without noting major concerns, and the statements or claims meet the above criteria, then the argument is more than likely sound and you can accept it as valid. Remember not to throw the baby out with the

Table 8.1. Critical reading checklist

Check for:	Notes
Clarity	If an argument is fuzzy, or unclear, it is probably poorly formulated – check if the argument is more clearly stated elsewhere in the text, or ask for a clarifying example, if you are engaging with a speaker. You may also need to Google a few terms if some of them are unfamiliar to you. If you have done this and the argument remains fuzzy, you should regard it as being less credible.
Accuracy	An argument can be clear, but not accurate. For example, 'The Earth is flat', is clearly stated, but inaccurate. Ask for, or find, verification or proof of claims from reliable and credible sources, even if the claims are simply and clearly written. Remember to check the sources of the evidence for credibility, as well as checking the general assumptions and claims made in these sources, as these relate directly to the overall accuracy of the claims. Also note whether the argument includes facts, opinions, theory, or faith – faith is not subject to proof, opinion should always be reinforced by other valid evidence, and theory must be well supported. If 'facts' are used, they should also be confirmed as such.
Precision	Details are important in claims. Instead of proving general statements, specific findings are important. For example, 'Corruption in South Africa has lost the country billions' might be true, but it is not precise. Besides not specifying a currency (rands or dollars?), it also gives us the impression that the author might have just repeated something they overheard – if, however, the author says that 'Global Financial Integrity, a prominent research group based in Washington, reported that South Africa lost R185-billion between 1994 and 2008 in illegal outflows (Corruption Watch, 2019)...', we may be more inclined to trust the claim: we have a source, and we have a specific amount, which shows that research has been done – as opposed to the 'news' comment-style writing of the earlier example.
Relevance	A statement or argument can be clear, and accurate, and precise, but it still might not be relevant. All supporting evidence must clearly relate to and support the main argument. For this, check the thesis statement, the topic sentences, and the supporting ideas. A sentence might look impressive, and even be well-researched, but if it does not support the topic sentence or thesis statement, it is irrelevant to the topic, and should be ignored.
Depth	If you have been asked to analyse something critically, it means that you need to get into the complexities and reasoning behind ideas. So, while simple theories are often the best ones, any theories or arguments used need to have been fully applied to the scenario under investigation, and not superficially mentioned or used as evidence without adequate explanation.
Breadth	Arguments should consider all relevant points of view – even when an author is trying to present one point of view, they need to show that they have at least considered other viewpoints to provide adequate breadth of argument. This demonstrates better objectivity, and may also show why the selected point of view is better than others. Other points of view may not necessarily be presented to as great a degree as the selected one, but if any of these leaves you asking why the author did not pursue this further, the writer has not done their job very well, and this should count against the argument.
Logic	The argument needs to be coherent throughout – individual statements and evidence need to add up to a unified, logical idea that convinces you, the reader. Identify claims that appear to contradict other 'evidence', and those claims that add seamlessly to the overall argument.

bathwater though – errors in logic can creep in. Even if one statement in a piece of writing is not completely sound, other elements of it may still have value and be credible. Use your SQ3R method to evaluate the piece of writing as a whole, and you will find it much easier deciding whether or not to use a particular source!

 For more on this, visit the Learning Zone and test your grasp of argumentation (see Chapter 8 Questions) and your ability to evaluate statements (Chapter 8 Evaluating statements). Both are online assessments.

 For lecturers, there is another activity on evaluating evidence available on the Learning Zone.

8.5 Recognising cognitive bias and logical fallacies

We now have a method to see if particular claims are 'safe' to rely on. But it may be more difficult to identify those that are not 'safe'.

One of the biggest challenges that we face as human beings is the resistance we get from our own brains: we simply loathe the idea that we can be wrong (Schulz, 2010). This means that we end up needing to deal with cognitive dissonance in a variety of ways – and not all of these are constructive.

While we *could* apologise, admit our mistake, and move on, we humans are more inclined to 'systematic deviations from logic' (Dobelli, 2013:2). We don't yet have a definitive answer on why we are this way (Ibid.), but perhaps we subconsciously suppose that we can maintain the appearance of being 'right' by using reasoning (even if it is faulty), or that we can delay the knowledge that we are wrong (Schulz, 2010).

Cognitive biases and logical fallacies are two important types of errors in logic. If you are honest as you read through Table 8.2, you are likely to recognise errors you yourself have made. Don't beat yourself up about these mistakes, but remember them so you can avoid repeating them in the future. Remember that it is essential that you can detect errors in logic in order to make valid judgements about arguments presented to you (although it is even more important that you recognise and try to avoid these errors when you are presenting an argument of your own!).

Okay, so what exactly are cognitive biases and logical fallacies?

A **cognitive bias** is the tendency to think in a systematically deviant way from sound or rational judgement (Dobelli, 2013). This bias is reflected in a pattern of logical errors committed over a period of time, and so does not refer to once-off errors in judgement. If you read your horoscope and think that a particular astrologer is always spot on, you might have suffered (or still be suffering!) the Forer effect: a cognitive bias that results in us attributing high accuracy to actually vague and general claims. For example, a claim like 'there is travel in your future' almost can't be wrong: whether you are travelling to class or going on an actual holiday, or 'travelling' now or in 20 years' time, that statement

Table 8.2. Common cognitive biases in argument

Name	What is it?	Example
Backfire Effect	*The tendency to strengthen your beliefs in the face of refuting evidence, likely because the evidence sparks a need to defend your identity.*	Global warming is a great example of this – despite a preponderance of evidence to support the theory of anthropocentric global warming, those who reject it sometimes become even more dogmatic that it is 'just a conspiracy'.
Bandwagon Effect	*Also referred to as the 'herd instinct' or social proof effect, we tend to agree with ideas and actions that the rest of the crowd follow.*	You decide to start the Banting diet because everyone else seems to be doing it, and it worked for one of your friends, even though you are not convinced that it is a healthy way of eating.
Bias Blind Spot	*We tend to assume that we are less biased than others, or identify more cognitive biases in others than in ourselves.*	Wondering why everyone else seems so guided by their emotions when you are able to put aside your feelings and previous ideas (in all likelihood, you are are also guided by your emotions).
Confirmation Bias	*We tend to seek, find and interpret information that confirms our ideas. The reality is that you are simply paying selective attention to the information that is available.*	For example, if you believe that left-handed people are more creative than right-handed people, you might keep finding evidence of this – because you already believe it, and you ignore evidence that right-handed people are creative too.
Dunning-Kruger Effect	*An effect in which incompetent people fail to realise their incompetence, and in which experts tend to underestimate their ability.*	Those ever-present 'organic' bloggers who advocate a 'chemical-free diet' are good examples of this – citing the dangers of dihydrogen monoxide (a fancy name for water or H_2O) is a sure way to prove you don't know what you are talking about!
Forer Effect	*Thinking that descriptions of personalities or predictions are more accurate and tailored to you than they really are.*	Telling your friend that the astrologer in the weekly paper 'knows their stuff': they said that your star sign was going to travel this week, and you are travelling to Durban on holiday.
Halo Effect	*The tendency to assume that because a person is good/bad at doing X, that they will auto-matically also be good/bad at Y and Z, i.e. evaluating positively (or negatively) everything a person does, whether this is true or not, because of known performance in one area.*	You visit a restaurant assuming that the chef, who is brilliant at cooking meat, will be equally good at preparing other kinds of food such as vegetables or dessert.
Illusory Correlation	*This is one of the most common errors. It assumes that there is a relationship between two events where there is actually no relation at all.*	You visited Paris once, and the people you interacted with were terribly rude. As a result, you assume that people in that city are ruder than elsewhere.

(continued)

Table 8.2. Continued

Name	What is it?	Example
Irrational Escalation	*Justifying further investment or belief in a decision because of previous investment, despite new evidence that suggests the belief is fault.*	Deciding to invest more money in a failing business venture because you've already invested so much time and money in it, despite the market being depressed and the product being a luxury item.
Ostrich Effect	*Deciding to ignore, rather than confront, a negative but obvious situation.*	Knowing that you have too many assignments, and too little time to get through them in the next two weeks, but simply carrying on with your normal social routine, pretending everything is fine.
Processing Difficulty Effect	*Surprisingly, information that is more complex and difficult to read and requires more thought is more easily remembered.*	When you are writing a test, you realise that the easy stuff that you thought you didn't need to focus on is completely gone, but that the really tough theory you struggled to grasp is all there – pity there is no question about it in the test!
Reactive Devaluation	*Deciding that an argument is worthless because it was written by someone you dislike or usually disagree with.*	In a group project, the person you like least in class is in charge – they came up with what you thought was an awful idea, and you wonder why everybody else in the group, and your lecturer, liked it.
Swimmer's Body Illusion	*We tend to take up products because of the appeal of their models or representatives, without realising that it is the models themselves that are appealing, and that the product likely has little effect on the model.*	Deciding to go to Harvard because it has a reputation as the 'best university', but not considering that it only recruits the best students, and so might only be the best university because its students are already the best – not because it offers the best teaching.
System Justification Bias	*People tend to defend the status quo in social, economic and political arenas – they reject and ridicule alternatives even if these alternatives would offer improvements.*	Any short scroll through the comments section of a controversial news article on Facebook will reveal a good example of this – those who offer constructive suggestions to problems often find themselves being abused on the basis of their race or gender for example.

The **status quo** means the existing state of affairs in society or politics. The phrase was originally Latin.

is just about impossible to prove wrong. However, it is not evidence of real knowledge about your future.

Table 8.2 above summarises some of the most common biases that you might encounter in your academic career. Take note of the specific examples and explanations of each to get a better idea of what the error in logic entails. The specific names are not very important, but the principle of the errors is important.

Nobody will expect you to remember all of these (although this is only a small sample of cognitive biases – we did warn you that human brains don't like to be wrong!), but being familiar with their existence does help you to identify faulty conclusions and assumptions in other people's arguments.

Use the online assessment Chapter 8 Identify the cognitive bias on the Learning Zone for more practice.

A **logical fallacy** is a concept closely related to a cognitive bias. It too reflects an error in logic. However, a logical fallacy is a mistake in reasoning that undermines the logic of an argument, and does not necessarily indicate a systematic pattern of behaviour – it can be committed as a single event, or as a series of mistakes.

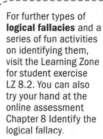

For further types of **logical fallacies** and a series of fun activities on identifying them, visit the Learning Zone for student exercise LZ 8.2. You can also try your hand at the online assessment Chapter 8 Identify the logical fallacy.

One of the most common logical fallacies, the *ad hominem* attack (in which the commentator attacks the arguer instead of the argument) is also abundantly evident on social and online media platforms. In fact, this is one of the reasons News24 shut down their comment section a few years ago! Many comments posted on these sites are illogical or fallacious through a lack of awareness, though some are deliberately provocative, and designed to create an argument unrelated to the topic. This is called 'trolling'.

'Trolling' comments use a variety of logical fallacies, some deliberate, some unintentional, to fan the flames. The best way to avoid falling into 'troll-traps' is to recognise them, and move on. Let us take a look at some of the most common logical fallacies that you might encounter:

- the **Ad hominem** attack: this fallacy uses an attack on the opponent's character to undermine the argument, instead of actually dealing with the argument itself. To avoid it, follow that old saying 'play the ball, not the player';
- **Anecdotal evidence**: This is one of the most common fallacies you will encounter, and most of us have used it at some point. When you use a personal experience or isolated example to 'prove' your claim, or dismiss statistically supported arguments, you have used an anecdotal event instead of actual evidence. For example, using your grandfather as evidence that smoking does not have a major impact on health because he smoked 40 cigarettes a day and is well at the age of 90 is anecdotal. Even if this is true, your grandfather's health certainly does not reflect the health of the majority of smokers;
- **Appeal to authority**: Claiming that someone in a position of power supports your position means that you have made a claim to authority. This is fine when that person is an expert in the field you are discussing, but it often happens in other areas – for example, celebrity Jenny McCarthy's position against vaccination (and her medically unsound claims that vaccinations cause autism) means that many people arguing against vaccines use her as an 'authority'.
- **Bandwagon**: using phrases like 'millions have tried' is a clear attempt to convince your audience by using popularity – however, just because millions of people have done something, does not mean that it is safe, logical, wise, or legitimate!;
- **Begging the question**: When someone uses the conclusion to an argument in the premise or statement itself, it is circular logic that

does not hold up as evidence. For example, saying that if actions were not illegal, then they would not be prohibited by the law. Simply concluding that the conclusion is true does not make it so. This is commonly seen in metaphysical statements and claims;

- **Black or white** fallacy: options seldom appear as exclusively black or white (e.g. completely right or completely wrong) – usually a number of alternatives also exist. The Black or White fallacy makes it seem as though only two options exist. A great example of this arose after the attacks on 9/11 (11 September 2001). Those who were against invading Iraq were labelled as traitors to the United States. The reality is that many people simply disagreed with the undemocratic methods used after this period, although they were loyal citizens of the United States;

- **Middle ground**: It might seem reasonable to try to find a compromise between opposing ideas, but sometimes one answer is right, and the other wrong. For example, after reading a convincing piece of literature by an expert immunologist that debunks the claim that vaccines cause autism, you decide that vaccines do still occasionally cause autism because there is so much other 'evidence' out there. Here, only one claim is supported by credible evidence, but you have used the middle ground fallacy and settled somewhere in the middle.

- **Personal incredulity**: Arguing that because something is difficult to believe it must be untrue is another fallacy – people disbelieve valid ideas all the time (for example, think back to Aristotle's battle to convince us that the Earth was a sphere);

> Jesse Richardson et al's (2019) poster 'thou shalt not commit logical fallacies', published under a Creative Commons licence which means it is free for you to use, explains many of these and more fallacies in a simple, easily understood way. It is available at <https://yourlogicalfallacyis.com/poster> We definitely recommend printing it out!

- the **Strawman**: when someone misrepresents an opponent's argument to make it easier to attack, they have used the strawman fallacy. This often happens in political debates; for example, if Politician X argues that we should focus our spending on education to improve outcomes for all, Politician Y claims that Politician X doesn't care about health spending;

- *Tu Toque*: This is also known as the 'you too' fallacy. Let's say, for example, that Speaker A is arguing against rampant consumerism in society. Speaker B tells Speaker A that he (Speaker B) can't really believe what she (Speaker A) is saying because she owns a cell phone, various appliances, and a fancy watch. Speaker B has committed the *tu toque* fallacy: although it may be hypocritical of Speaker A to argue against consumerism while owning many consumer goods, her argument is not invalid because she does own these goods;

You have likely already noticed logical fallacies or cognitive biases in your own arguments or in those of other people. As a critical reader and scholar, try to be aware of these mistakes so that you can recognise faulty argumentation and make informed decisions about the validity of arguments.

 Use the student exercise LZ 8.3 on the Learning Zone to reflect on your own cognition more formally.

8.6 Conclusion

This chapter highlighted the importance of adopting a critical lens in academia, and offered a practical method of critical analysis to ensure that you consider arguments objectively. By examining concepts related to 'truth', various influences on human decision-making, and a number of cognitive biases and logical fallacies, it also laid the foundation for Chapter 9, which focuses on using an awareness of these pitfalls to develop logical, supported claims.

References

Bhaskar, R. 2008. *A realist theory of science*. London and New York, NY: Routledge.

Burbules, N. C. and Berk, R. 1999. 'Critical thinking and critical pedagogy: relations, differences, and limits'. *Media Education*. Retrieved 30/08/2019, available at <http://mediaeducation.org.mt/wp-content/uploads/2013/05/Critical-Thinking-and-Critical-Pedagogy.pdf>

Corruption Watch. 2019. 'What's the real cost of corruption: part one'. *Corruption Watch*. Retrieved 18/11/2019, available at <https://www.corruptionwatch.org.za/whats-the-real-cost-of-corruption-part-one/>

Descartes, R. 2009. *The meditations, and selections from the principles of René Descartes (1596–1650)*. Charleston, SC: BiblioBazaar

Dobelli, R. 2013. *The art of thinking clearly*. London: Sceptre.

Du Plooy-Cilliers, F., Davis, C., and Bezuidenhout, R. 2014. *Research matters*. Cape Town: Juta.

Krauss, S. E. 2005. 'Research paradigms and meaning making: A primer'. *The qualitative report*, 10(4): 758-770. Retrieved 29/08/2019, available at <https://nsuworks.nova.edu/tqr/vol10/iss4/7/>

Lilienfeld, S. O., Lynn, S. J., Ruscio, J. and Beyerstein, B. L. 2010. *50 Great myths of popular psychology: Shattering widespread misconceptions about human behaviour*. Chichester: Wiley-Blackwell.

Makhmaltchi, M., Sajjadi, S. M., Noaparast, K. B. and Fardanesh, H. 2017. 'Disciplinary lenses model: A new approach to collegiate-level general education'. *Educational research and reviews*, 12(23): 1129-1137. [Online]. Retrieved 29/08/2019, available at <https://files.eric.ed.gov/fulltext/EJ1163296.pdf>

McLeod, S. 2018. 'Cognitive dissonance'. *Simply Psychology*. Retrieved 28/08/2019, available at <http://www.simplypsychology.org/cognitive-dissonance.html>

McLeod, S. 2010. 'Perceptual set'. *Simply Psychology*. Retrieved 24/08/2019, available at <https://www.simplypsychology.org/perceptual-set.html>

Nielsen, J. A., Zielinski, B. A., Ferguson, M. A., Lainhart, J. E. and Anderson, J. S. 2013. 'An evaluation of the left-brain vs right-brain hypothesis with resting state functional connectivity magnetic resonance imaging'. *PLoS ONE*, 8(8). Retrieved 30/08/2019, available at <https://journals.plos.org/plosone/article?id=10.1371/journal.pone.0071275>

Professor Dave Explains. 2018. 'No, you're not left-brained or right-brained.' *YouTube*. Retrieved 26/08/2019, available at <https://www.youtube.com/watch?v=mKlkwJypzJQ>

Richardson, J., Smith, A., Meaden, S. and Flip Creative. 2019. 'Thou shalt not commit logical fallacies' (poster). *Your Logical Fallacy Is*. Retrieved 01/09/2019, available at <https://yourlogicalfallacyis.com/poster>

Ross, L. and Ward, A. 1996. 'Naive realism in everyday life: Implications for social conflict and misunderstanding.' In Reed, E. S, Turiel, E., and Brown, T. (Eds.) 1996. The Jean Piaget Symposia Series. *Values and knowledge*: 103-135. Hoboken, NJ: Taylor and Francis.

Schulz, K. 2010. *Being wrong: Adventures in the margin of error*. London: Portobello Books.

CHAPTER 9

Problems, persuasion and argument

'...[Be a] lion, roaring in the forests of knowledge and wisdom...'
– 'Abdu'l-Bahá (2020)

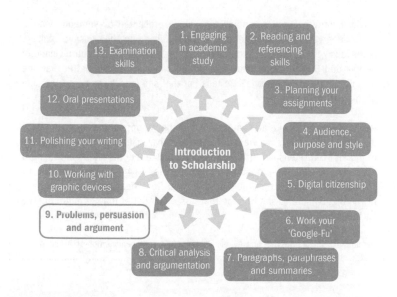

- 1. Engaging in academic study
- 2. Reading and referencing skills
- 13. Examination skills
- 12. Oral presentations
- 3. Planning your assignments
- 11. Polishing your writing
- **Introduction to Scholarship**
- 4. Audience, purpose and style
- 10. Working with graphic devices
- 5. Digital citizenship
- 9. Problems, persuasion and argument
- 6. Work your 'Google-Fu'
- 8. Critical analysis and argumentation
- 7. Paragraphs, paraphrases and summaries

OBJECTIVES

At the end of this chapter, you should be able to:
- understand the importance of using sound, reasoned arguments in an academic context;
- be familiar with the nature of problems and their characteristic features;
- recognise the role that mental models play in shaping our approach to problems;
- be able to use suitable combinations of the Toulmin method and Aristotle's three rhetorical appeals to develop solid arguments;
- understand the differences between inductive and deductive reasoning;
- use GASCAP forms of argument appropriately in your writing;
- be able to develop sound, reasoned arguments in both short and extended pieces of writing;
- use given tools to reflect on the quality of your argument.

Refilwe is sitting at the university coffee shop, but instead of relaxing, her face is set into a tight scowl -- she is clearly hopping mad. When her friend David arrives, he notices her mood immediately and asks her what she is so upset about. She erupts, telling him that her philosophy lecturer has given her group the task of supporting the utilitarian view that the act of killing one child to save a village of people is *justified*! She argues that as a volunteer at Childline, she cannot, and simply *will* not, defend the decision to kill an innocent child – for any reason. She punctuates her angry tale by telling David that her 'ridiculous lecturer' should have placed her in the opposing team instead.

David (clearly not understanding the danger presented by someone who is deeply angry) tells Refilwe that she is crazy. Killing one child to save a village is obviously horrible, but of course it is better than letting an entire village of people die! Surely the decision that protects more people is better than the decision that saves just one? He tells her that she is lucky to be on the side of the **utilitarians**.

Refilwe respects David, and she finds his response both surprising and shocking. When she rejects his view and expresses her disappointment at his 'lack of morals', David tells her that she needs to think about the situation rationally instead of reacting so emotionally (careful there, Dave!) He reassures her by saying that he agrees that the situation is far from ideal, but, if she can look at the bigger picture, she will see that it is much better than letting an entire village of people die.

Disclaimer: Please note that no Davids were harmed during the invention of this scenario.

A **utilitarian** view argues that the consequences of an action are important, and that, in deciding what is right and wrong, one should consider the safety and happiness of the greatest number of people, above those of any individuals.

9.1 Introduction

The horribly unpleasant scenario that Refilwe has been told to defend is an extreme one, but is quite typical of a situation that philosophy, psychology, law or sociology lecturers may ask students to try to resolve. This one may seem far-fetched, but this kind of scenario does exist in real life.

For example, you might recall the moral dilemma that Corrine Rey, a cartoonist at the French *Charlie Hebdo* magazine, faced in January 2015. The cartoonist and her daughter were threatened by two gunmen at the entrance to the magazine, where she was told to enter her key at the door's security pad. Corinne agreed, and the gunmen spared her and her daughter. Once they got inside the office though, the gunmen killed twelve other people (Alderman, 2015).

Rey's actions sparked mixed reactions, with a few critics arguing that she should have sacrificed her daughter and herself to save the other employees of the magazine: after all, she must have *known* that the gunmen were there to kill.

Is this a fair criticism?

Think about the utilitarian argument in the earlier IRL scenario, and then try to apply it to Rey's decision. Looked at from this perspective, many of us might agree with this criticism and argue that we would have sacrificed ourselves if we had been in that position. To defend *this* claim, we could say (and David from our IRL scenario would agree) that sacrificing two lives to save twelve lives in the office upstairs makes logical sense, and so argue for sacrifice as the best course of action.

Unfortunately, the argument is not quite as simple as that. For one thing, it assumes that there are only two courses of action, that is, to key in the pin or to not key in the pin, and only two possible results, that is death for two outside the office, or death for twelve inside the office. In reality, there are seldom such black or white, binary decisions, and this is the case here too. Other courses of action (like screaming for help, or running away) may have returned different results too.

The second problem with the argument is that it completely ignores the fact that most life on this planet has an innate, or inborn, instinct to survive; in mammals, an even *stronger* instinct to protect offspring exists. In short, we are genetically programmed to live and to increase the likelihood of our genetic survival by protecting our children. From a biological perspective at least, Rey's actions may not be so illogical after all.

Finally, the argument also fails to consider contextual factors. In this case, the publication had received many death threats, and so we might counter-argue that the management of the magazine should have ensured that the building was better secured or guarded. As an employer, the magazine had a duty to ensure that its premises were safe for its employees. From this perspective, Corinne Rey had a right to expect a safe place to work, and so should not be criticised for her actions at all.

For more examples of similar dilemmas, visit Kelly Ross's page about moral dilemmas, available at <http://www.friesian.com/valley/dilemmas.htm> and think about what you would do in these scenarios, and how you would defend your choice.

We have just performed a very simple critique of an argument using the *Charlie Hebdo* example. Without any formal training, we analysed a rather complex problem and began to dissect it, looking at it from different perspectives, and deciding which assumptions were valid, and which were problematic. At the end, we came to a form of conclusion that a simple utilitarian approach was not quite good enough to support the criticism against Rey, and that we needed to consider other elements.

We hope *you* will never have to face a real-life situation that has exclusively terrible alternatives, and that any 'impossible' decisions you are asked to take or defend are of the hypothetical, academic kind. However, even these hypothetical problems can be highly annoying, or seemingly impossible to resolve, as Refilwe has discovered.

Asking you to defend or oppose particular decisions or actions is a very effective way of teaching you to think more critically, helping you produce sound, rational arguments. But what *is* a rational argument? And why *should* our arguments be rational? And, even assuming that they *do* need to be rational, how are we supposed to separate our strong beliefs from our arguments?

Rhetoric refers to specific language or literary devices such as figures of speech that are used in writing to persuade or impress an audience. Martin Luther King's famous speech, in which he uses the phrase 'I have a dream' nine times, is a fine example of the successful use of a rhetorical device.

This chapter will explore these questions by looking at some of the typical features of complex problems, and explaining one of the ways in which our brains deal with them. The chapter will then discuss the use of a particular method of argument, alongside three forms of rhetorical appeals that work together to build your arguments. These discuss different types of reasoning, ensuring that you establish credibility as a writer, and using emotional appeals and rhetorical devices in your writing.

By guiding you through the use of these strategies, you will get to grips with the 'how' of developing and presenting coherent arguments and evidence in a convincing, academically sound manner. To kick us off then, let us examine the nature of the problem itself.

9.2 Understanding problems

A **problem** is a question or situation that involves an element of doubt, uncertainty or difficulty. However, it can also refer to a question that is proposed for discussion – as in Refilwe's earlier debate – or to a mathematical statement that requires a solution, usually a relatively simple one (Hornby, 2010), although that depends on how good you are at maths.

In academic contexts, you will be asked to discuss a variety of problems and will probably have to deal most frequently with those that do not have easy, clear-cut solutions. Although it might be easy to solve the 'problem' of what to wear to class, problems like poverty are much more complex to define, let alone solve. Importantly, understanding and dealing with problems requires objective, critical thought.

As in Refilwe's case above, there is a good chance that during the course of your studies you will be exposed to ideas that make you angry, that make you question long-held beliefs, or that leave you feeling deeply uncomfortable. This may not sound like much fun, but challenging existing ideas and beliefs is a critical component of developing as a scholar, and, more importantly, is one of the important ways in which we progress as a society.

Throughout your studies, keep in mind Oscar Wilde's assertion that '[a]n idea that is not dangerous is unworthy of being called an idea at all' (Wilde, 1997:181) – which echoes the introductory quote for this chapter advising you to be a lion 'roaring in the forest of knowledge'. The gist of these quotes is that you do *not* need to accept current beliefs about problems like a timid little lamb – as long as you can provide sound, coherent evidence against them. You are also free – hopefully even encouraged by your lecturers – to develop *different* approaches or solutions to problems, if you have clear and logical evidence to support them. But to be in a position to challenge existing beliefs and ideas, you need a better idea of how complex problems are dealt with by the human brain.

9.2.1 Complex problems and modelling

Complex problems exist in all spheres of life – from treating medical problems or deciding on the best economic policy to decrease a wealth-gap, to finding the most appropriate location and time to complete a group assignment. Many problems might be complex in that they include a number of different elements, but they are not necessarily all equally *difficult* to resolve.

Spector and Park (2012) do, however, argue that most complex problems have:

- both cognitive and non-cognitive (factual/knowledge) elements;
- interrelated aspects that influence each other;
- multiple routes to solutions.

In addition to these three points, *previous experience with similar problems* is a fourth and very important one in dealing with new problems (Ibid.). This is because humans build and use **mental models**, or frameworks, for analysing, understanding, simplifying and resolving problems. These models are constructed through various, unique learning experiences; depending on the problem and required outcome, different models will be selected for use.

In other words, when we encounter a 'problem', we automatically try to determine if it is in any way similar to previous problems. If it is, we can use an existing model to decide how to proceed in the new situation, and adapt the existing model to our new problem in order to find a solution. This is why practising previous examination papers helps you to prepare for upcoming exams. By practising and getting familiar with the format and types of questions, your brain develops suitable models for dealing with the different problems – for new questions, you then just adapt these mental models (Bernstein, 1990).

Obviously, in situations in which you have no previous experience, a problem will require much more effort than one that is familiar. For example, if you were studying English literature and someone gave you an advanced statistical equation to complete, it is highly likely that you would be unable to figure out the equation without a substantial amount of research and/or assistance, even if you did complete mathematics at school. On the other hand, if you were studying third-year Econometrics, you might find this task a challenge, but because you already have access to equation-related mental models, you should be able to figure it out after some thought and application.

Mental models are very useful and prevent us from reinventing the proverbial wheel (imagine having to learn how to drive every time you got into a new vehicle!), but relying on existing models does mean that we tend to stay 'safe' when we solve problems. Instead of trying to think of a way of solving a problem in a completely different way, we tend to apply existing models instead (Spector and Park, 2012; Bernstein, 1990). This works in many cases (we would not have developed this adaptation if it didn't), but sometimes we do need a completely fresh approach to solve something – and, unfortunately, fresh approaches (the ones that Oscar Wilde admiringly calls 'dangerous') usually come with a serious mental workout!

9.2.2 A practical example of modelling in action

Let us look at our earlier example of Refilwe's problem to explain the previous section a little more clearly. In the IRL situation, Refilwe has more than just the problem of trying to prove that a utilitarian approach is a defensible option – and this is a complex, and refutable, argument in and of itself. The more pressing problem (since it needs to be resolved before she can start to generate reasonable arguments) is her current outright rejection of the idea of killing a child, regardless of what the action achieves.

If we relate Refilwe's problem to the concept of mental models, we could say that she is not familiar with utilitarian theory, or has only recently been exposed to it, so she does not yet have a mental model that includes these theoretical concepts.

However, there are many situations, in real life, in movies, and in books, in which the sacrifice of one helps to save many, and Refilwe may indeed have a mental framework that includes examples of *these* situations. So, instead of trying to develop a brand new mental model (pretty hard work), she could apply her new theoretical knowledge of utilitarianism into her existing practical framework instead. If she can manage this, she may just be able to overcome her complete refusal to argue for the approach.

But Refilwe isn't making this leap herself. Can David help her to make this leap instead?

Absolutely – in fact, this is often the way that lecturers or teachers manage to evoke those 'aha' moments from students: by providing examples of similar situations, an appropriate, pre-existing mental model is triggered into action (Bernstein, 1990), and the student may better comprehend an idea that was previously too abstract or foreign to understand.

In Refilwe's case, David's best chance of persuading her may be to relate a particular instance, real or fictional, in which one person was sacrificed for the benefit or safety of others (rather than telling her not to be so emotional!). In particular, bringing up a *shared* example that both he and Refilwe are familiar with may prove especially useful for this purpose.

If David can come up with such an example, Refilwe's brain can then access the example, which should be 'filed' with other similar examples in an existing mental model. Once this is located, the process of associating utilitarian theory with instances linked to her existing mental model may begin (Bernstein, 1990). And this new, more positive association may allow Refilwe to think more objectively about the problem and help her to engage with the task, even if it is unlikely that anyone will be able to remove her abhorrence of the sacrifice of a child. In short, Refilwe can use an existing mental model as a framework, and adapt it to accommodate the new information. In the process, she can begin to solve the first problem, so that she can reason more formally about the second one.

To be an effective problem-solver, you need to be able to manage broad chunks of knowledge, and also be able to perform critical thinking activities like analysis, synthesis and evaluation of both new and existing information. Perhaps most importantly, to be able to reason effectively as a problem-solver, you also need to know when and how to use and adapt existing mental models, or be prepared to try a completely new approach to problem-solving, when this is required (Spector and Park, 2012).

The next section outlines a method that you can use to arrange your arguments in a methodical and convincing way. You will notice that you are already familiar with some elements of this structure and process, so try to use your existing understanding of argumentation as a framework, and add any new understanding to this.

9.3 Sound arguments

No matter what faculty you are studying in, as a tertiary student, you will inevitably be required to analyse other people's arguments and to develop your own. In fact, you already argue with your friends and family in an informal way, and will have argued at school in a slightly more formal way when you wrote essays. But how did you learn to argue? Can you recognise a bogus argument from a sincere one? Do you really know what an argument is? Or did teachers just more or less expect you to figure it out on your own?

Because arguing is such a common, everyday phenomenon, we may mistakenly assume that being good at arguing will come naturally. As a result, it is quite possible that you have received little formal instruction in the art of arguing. If you have frequent informal arguments with family or friends, yes, you might have had a bit more practice. But the reality is that very few students arrive at university with a clear idea of what a good formal argument looks like, or how to go about constructing one.

You might agree, then, that before we try to build formal arguments, we need to take a step back and first really understand what constitutes a formal argument, and what structural elements make it up (Larson et al., 2009).

Simply put, a **formal argument** is an attempt, through use of a main claim and supporting evidence, to persuade readers or listeners to change their minds (or behaviours) on a particular topic by providing clear and credible reasons (Ibid.).

9.3.1 The Toulmin argument structure

Stephen Toulmin, a British philosopher, argued (2003) that well-structured practical arguments will include six different elements. Using a hypothetical example and 'evidence' to aid your understanding, these six points include:

- **A claim**: this is a disputable statement, usually your thesis statement, and the point that you are trying to get your reader or audience to accept (e.g. *Completing previous examination papers helps many students pass their examinations* – you are going to try to convince your reader of this point. You have not yet provided evidence in support of it though.)
- **Data/Evidence**: this refers to the (hopefully) credible evidence that you offer to support the claim (e.g. *A study conducted by Smith and Walton (2019) found, for example, that ....* This is the kind of evidence with which you can support your claim – recent studies in the field that back your claim are obviously particularly useful, but other kinds of data might be useful too.)
- **A warrant**: this is the assumption, usually unstated, on which the claim and evidence depend (e.g. the assumption is *that it is desirable to be well-prepared for exams so that you can perform better* – this is not specifically stated, but is implied in the claim and supporting evidence.)
- **Backing**: this refers to support or justification that the *warrant* itself is true. Including this shows that you have not only thought about the claim, but also fully analysed the assumptions that you have made about the topic (e.g. *Successful performance at university level is closely associated with graduates' chances of finding jobs and in determining levels of job satisfaction [see Reed, 2017; Smythe and Fredman, 2017; and French et al., 2015].*)
- **Qualifier**: this is a statement about how strong the claim is; it can therefore either limit or enhance the degree of certainty and scope of your claim (in the example, the claim includes the word 'many', meaning that it does not try to argue that this is true for *all* students – this makes the statement more believable than if we had said 'all students', or just 'students'.) Academic language should be cautious ('many'), and not absolute ('all'). As a result, the statement uses hedging to avoid absolutes in arguments. This reduces the degree of force and makes it less likely that your claim will be refuted since being absolute is often a red flag to someone analysing your argument.
- **Rebuttal/reservations**: this is an exception to the claim. If you acknowledge that your claim does not always apply, or you include counterclaims, you are likely to be taken more seriously since you are demonstrating that you have looked at your topic from different perspectives. You should do more than simply include counterarguments though – you should deal with them too.

In academic writing, **hedging** means including cautious statements that reduce the certainty of a claim or that help to express a statement more politely. For example, instead of claiming: 'The evidence shows that X has a direct impact on Y', it is more academically suitable to say: 'The evidence suggests that X has a direct impact on Y'.

Werry (2019) suggests three ways to deal with counterarguments: you can either make a *strategic concession* by acknowledging that a different view has merit, and then include some parts of it in your own argument; or, you can *refute* the counterargument by revealing its shortcomings; or, you can show that the counterargument is *irrelevant*, or not quite relevant to your particular claim.

Toulmin (2003) says that the first three of these six elements are essential, while the second three are not *always* necessary. If you want to be as thorough as possible, try to use all six, if possible. You should be able to decide whether or not the second three elements are necessary after examining your particular claim, its unstated assumptions, and the evidence you have to support your claim and warrant.

If you decide to skip any of these non-essential three elements, make sure that you are not simply avoiding any valid counterarguments – choosing to ignore valid concerns about your argument may mean your claim is incorrect, and your reader or audience may dismiss it. However, if you can acknowledge that a counterclaim exists and then dismiss it in one of the ways that Werry (2019) suggests, you will strengthen your own argument, and your audience should feel safer in trusting your claim.

The Toulmin method provides you with the six core 'ingredients' of a practical or formal argument, but it does not actually provide you with the tools to frame these components. Nor does it explain the reasoning behind these elements. This next section outlines three rhetorical strategies that are useful, in varying combinations, in developing your argument.

9.3.2 Using rhetorical strategies for persuasion

In previous chapters, we covered some essential areas of planning your writing. These included the importance of developing strong thesis statements for your essays and arguments (Chapter 3), doing research to find suitable and credible sources (Chapters 3 and 6), writing for your audience (Chapter 4), and using a suitable skeleton framework to organise your essay (Chapter 3).

When you are developing an argument, it is vital that you do not forget about these. In addition, you need to ensure that your argument is logical, credible, and appealing to your audience. These particular criteria can be achieved through the use of three rhetorical strategies.

Okay, but um … just what is rhetoric anyway?

When we refer to **rhetoric** in the context of developing arguments, we refer to the skilled use of language for the purpose of persuasive communication (Hornby, 2010). Rhetoric was developed in ancient Italy and Greece, and includes a number of strategies and literary devices that help to make the arrangement, style, and delivery of arguments more convincing – if used appropriately.

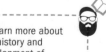

To learn more about the history and development of rhetoric, we recommend Richard Toye's work *Rhetoric: A Very Short Introduction*. 2013. Oxford: Oxford University Press.

Because of its origins, you will find that many of rhetoric's terms have Greek or Latin origins, so they may look like, well … Greek. Don't worry about the terminology now. We introduce a little terminology here, but for your purposes, the importance of rhetoric lies in *applying it*, not in recognising its terminology.

As you work your way through these strategies and devices, you may notice that the devices themselves are recognisable in speeches or writing that you are already know. This shows that rhetoric is not an entirely new practice to you, and that you encounter it, and even use it, very often in your everyday life (like Refilwe did in our earlier IRL scenario). Because of this, you might be able to attach a few of these practices to a list of examples for which you have already created a mental model. When you work through these strategies, try to focus on understanding the principles behind them, so that you can use these strategies effectively in your own writing.

Aristotle (350 **BCE**) argued that three particular appeals are available to someone delivering an argument, for their argument to be persuasive. The first option is to prove the truth of a statement, the second is to convince the audience of the person's honourable character, and the third is to appeal to the audience through their emotions. In rhetorical terms, these three options are called *logos* (logic), *ethos* (credibility), and *pathos* (emotion), and you can use each in varying degrees, depending on what you are trying to achieve with your argument. The section below deals with each of these appeals, explaining how they may benefit your arguments.

Note!

BCE stands for 'Before the Common Era', while CE stands for 'Common Era'. Using BCE and CE is a newer, more academically accepted form of dating than the practice of using the older BC (Before Christ) or AD (Anno Domini) forms. Dates using the new system are, however, numerically equivalent to the old system – 400BCE is the same as 400BC, and 2016CE is the same as 2016AD – so you shouldn't have any problem making the switch, especially since neither system has a Year 0.

Logos

Logos, or logic, is the formal system of analysis that uses reason to demonstrate the reliability of a claim. In its purest form, it tests statements in a logical sequence, and arrives at an indisputable conclusion. Aristotle (350 BCE) developed what is known as a **syllogism** to describe one such logical sequence. In it, he offers two premises and arrives at a conclusion based on these premises:

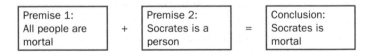

| Premise 1: All people are mortal | + | Premise 2: Socrates is a person | = | Conclusion: Socrates is mortal |

In the syllogism, if both premises are accepted as true, then the conclusion naturally follows as true too. In this example, we have evidence that people are mortal, and that Socrates was, indeed, a person. We can therefore deduce that the conclusion (that Socrates is mortal) is valid. This follows an 'if A = B; and B = C; then A=C' formula, and the particular order and logic of the premises is key to determining the validity of the statement.

This type of reasoning is called **deductive reasoning**. To be valid, deductive reasoning requires that an argument has a great deal of reliable evidence, and then argues from the **general to the specific**.

For example, we could argue that chimpanzees are classified as mammals, that all mammals (as evidenced by numerous medical and zoological evidence) have kidneys, and so all chimpanzees have kidneys.

If either one of these premises were proven to be false, the conclusion would be invalid. But as long as you can support the two premises, you can safely conclude that chimpanzees do, indeed, have kidneys. If you use this form of argument, ensure that you place your premise items in the correct order to ensure logic.

Deductive reasoning is very appealing because of its apparently faultless logic, but problems seldom present themselves as such easily resolved concepts. Instead, multiple premises with multiple warrants and conflicting evidence are often involved. To help us work towards logical argument in more complex cases, we often require the use of inductive reasoning.

In **inductive reasoning**, an argument takes a particular example, and then tries to move towards generalisations or conclusions based on that one example, or a smaller set of examples. In other words, it moves from the **specific to the general**. If you think about this, you might realise that statements of inductive reasoning will not have as much validity as those made by deductive reasoning. For example, think about the caution against using anecdotal evidence to draw conclusions, which we raised in Chapter 8 of this book. It would be incorrect to conclude that, because something happened once to one person, the same thing, or similar things, will happen to a larger group of people.

It is possible to improve the validity of inductive reasoning, but to do so often requires a number of extra steps or elements. Some of the more common argumentation strategies that you could use with this method include the six GASCAP forms (Werry, 2019), which are listed below (when using these, you will need to ensure that you do not commit logical fallacies (see Chapter 8)):

1. **Generalisation**: This is a very common form of reasoning in research. To be reliable, it must use a carefully selected and representative sample group to draw conclusions about a larger, but similar population.
2. **Analogy**: In this form, you draw on similarities between one situation and another, and try to make inferences about the one situation from knowledge about the other. This is seldom convincing in itself. To be considered acceptable, it is essential that these situations be similar in context and nature – that is, you need to be sure that you are comparing 'apples with apples' – and that you are not simply drawing on anecdotal evidence. Legal precedents are reasonably good examples of this form of reasoning, since actions in one situation, for one crime, offer a good example for similar, but later, situations.
3. **Sign**: This form of reasoning assumes that with the presence of one sign, another will also be present. For example, people might assume that you will be successful in your career because you do

well in your studies, or vice versa. In many cases, this might be true, but we cannot reliably use signs to make predictions; if you are going to use this form of logic, it would be wise to apply some other forms too.

4. **Causality**: Although arguments often try to use the 'if X, then Y' argument structure, it is really important that you are sure that X does actually cause Y. It often happens that certain things have a correlational relationship (that is, they appear together). However, this does not mean that one causes the other. The first danger is making a false assumption that X causes Y. If a causal relationship is seen, the second danger is determining which factor caused the other, as this can be difficult to determine. A silly, but illustrative, example is that reports of theft and consumption of ice cream both rise in summer. Although both of these factors increase at a particular point, this does not mean that ice cream sales cause theft, or that theft improves ice cream sales. Instead, the factors have little to do with each other. Try to steer clear of this form of reasoning, unless a scientific study has demonstrated a causal relationship between two elements you are using in an argument.

5. **Authority**: This reasoning relies on the credibility of your source. Generally, the more expert your source is, the more easily you will be able to convince your audience that your claim is valid. Again, watch out for the logical fallacy of appealing to authority in which you try to use famous people to prove your arguments – in academic arguments, experts in a particular subject area who have credible research and academic credentials are very valuable supports. Celebrities, however much they may speak about an issue, are generally not!

6. **Principle**: The final form of reasoning that you might want to use is an appeal to a principle that is widely accepted as valid. Appeals to well-supported theories (such as evolution or anthropogenic climate change) are usually a good way to support your argument or a place on which to base any warrants. If you do use this, just ensure that you are familiar with any counterclaims and arguments against the principle, and deal with these, if necessary, in your reasoning.

In addition, ensuring that you have not committed any logical fallacies or engaged in any cognitive bias is a significant factor in developing reasonable, logical arguments. For this reason, refer back to Chapter 8 when you analyse your arguments to ensure that you have not erred in this way.

Ethos

Ethos, as one of the rhetorical appeals, refers to the 'ethics' or character of the presenter of the argument (you). In short, in order to 'sell' your argument, it is important that your audience sees you as believable and trustworthy. We dealt with credibility in Chapters 4 and 6; to ensure

Point of interest

There are some celebrities with impressive academic credentials. Pitch Perfect star Rebel Wilson has a degree in law; 'Big Bang Theory' actress Mayim Bialik has a PhD in neuroscience; comedian Ken Jeong is a medical doctor; and Lupita Nyong'o has a Master's degree in acting from Yale! However, if you wanted to reference them, you would need to reference their academic research papers, rather than media statements or tabloid reports.

that you are seen as credible, check that you have followed the advice in both of those chapters. There are a few additional tips that you can follow to improve your audience's perception of you:

- Make sure that you come across as objective by including counterarguments – this tells your reader that you have at least considered opposing viewpoints;
- Remember to write for your readers, or speak to your audience in a way that they will understand;
- Disclose any element of personal gain or interest in the topic that you might have (again, if your audience is reading your argument about how wine is healthy, it will help them to put things in context if they know that you work at a wine farm);
- Make sure that you revise and proofread your argument carefully so that it follows a logical structure and that sloppy mistakes don't cast doubts on your credibility.

Pathos

Aristotle's third rhetorical appeal, **pathos**, refers to using rhetorical devices to appeal to your audience's emotions. This form of persuasion can be seen as manipulative, and may be viewed negatively. For example, it is often used by advertisers and politicians when they are trying to sell us products, or convince us to vote them into power.

However, if it is used carefully and appropriately, it can serve a legitimate purpose. Although logic is usually effective in persuading people to do something, stubbornness occasionally sets in, and then logic is simply not enough. A good example of this is how long it has taken many people to accept the theory that humans have played a significant role in accelerating climate change, despite the huge volume of scientific evidence in support of it. As a result, action to prevent further damage has been delayed by the public.

In cases like this, relying on more personal appeals, instead of arguing logically, might be more effective. For example, those arguing for more urgent action on climate change might tell of one person's thoughts and concerns about their young child growing up in a world lacking fresh water or food, in the form of an advertisement, and find that this form of persuasion has more of an effect on some of their audience than a logical format did.

Appeals to emotion can also be useful in certain news stories, in which you are trying to raise awareness of a particular societal concern. But if you *do* make use of an emotion appeal, it is really important that you do not use it to misrepresent your case, or instil fear (like adverts for security doors often do in South Africa!). You should be safe using an emotional appeal if it can support your claim, or show its relevance, but make sure that you are not trying to manipulate your audience. Emotional appeals also risk being subjective because you frame the argument based on your beliefs, and not from reasoned argument. When you rely on your beliefs, and are unable to come across as impartial or credible, there is little chance that your audience will buy into your argument.

An outline of rhetorical strategies is not complete without mentioning some of the more common rhetorical devices that you might encounter in other arguments or make use of yourself.

Specific rhetorical devices

An important part of rhetoric is the particular literary and writing techniques that are used to persuade an audience about a particular argument. This section covers some of the more commonly used, helpful examples that you may wish to use, as well as some problematic ones that you may find in the arguments of others. Chapter 11, which deals with polishing your writing, covers some of these in more detail, but a brief outline of them under this section is useful as they are clearly part of the tools of argument.

For a more detailed list and description of rhetorical devices, together with clear examples, visit the Learning Zone and review the student resources for this chapter (including student exercise LZ 9.1, and online assessments on GASCAP forms and Ethos, logos and pathos).

Table 9.1. Common rhetorical devices in argumentative writing

Rhetorical device	Description
Amplification	By describing a particular concept in more detail, you emphasise your point, and help the audience to understand your argument.
Anaphora	Anaphora is the repetition of words or phrases at the start of a sequence of sentences. Martin Luther King's 'I have a dream' speech is a good example.
Anticipated objection	With this approach, the writer anticipates, or pre-empts, counterarguments, and responds to them, even though the reader may not yet have voiced these objections. If done well, it lets your readers think that you have thought of everything, but if done poorly, your readers might feel that you are putting words into their mouths, especially if you do not provide correct context or credible counterarguments.
Antithesis	This involves placing two opposing ideas together in a sentence to achieve a contrasting, or highlighting, effect. You may recognise antithetical terms or quotations like 'bittersweet', 'Give every man thy ear, but few thy voice' (Shakespeare, 1998: *Hamlet* 1.3.65) or 'It was the best of times, it was the worst of times, it was the age of wisdom, it was the age of foolishness…' (Dickens, 1981: *A tale of two cities* 1). You will note that each side of the statement uses a similar structure, which highlights the contrast.
Argument by analysis	If you break up the subject matter into its component parts, and use these as the structure for your argument, then you are using argument by analysis. This is as useful in the natural sciences as it is in philosophical arguments, but be careful to keep the 'bigger' meaning in mind if you use this approach, as it is easy to get lost in the details.

Table 9.1. Continued

Rhetorical device	Description
Circumlocution	This tongue-twisting Italian word refers to an argument that circles around an idea by using many vague words, instead of getting to the point. This can be because of deliberate euphemism (talk about unpleasant things in a roundabout way that makes them seem less harsh), or because of a deliberate muddying of meaning (you may have noticed that politicians often use this when responding to difficult questions). Usually, however, circumlocution happens because arguments have been poorly proofread or edited. We will discuss this more in Chapter 11.
Connotation	By using the implied meaning of a word, or its associations, rather than its primary or straight meaning, a writer can easily convey emotions. For example, you could use the word 'home' instead of the word 'house'; 'home' implies emotional warmth, family and comfort, which makes the audience engage with it more than the word 'house' which suggests a building without an emotional connection.
Parallelism	An important rhetorical tool, 'parallelism' refers to ensuring that your writing form remains mirrored, consistent, and balanced. Words, phrases or clauses that have a similar sound or structure are carefully combined to create a sense of balance and elegance for the reader. This applies, for example, to verb forms (e.g. -ing or -ed forms), but also to the types of clauses and sentences used. We will deal with this in more detail in Chapter 11 as it is an important feature of writing style.
Parenthesis	Including extra information within two commas, two brackets, or two dashes is a relatively neat way to provide context, or information that might help to convince your audience. It can also be used to inject opinion or contrast, depending on the formality and context of your writing.

9.4 Checking your argument

This chapter offered two main tools to help shape your arguments: the Toulmin method, and Aristotle's three rhetorical appeals. These are both very useful for evaluating the outcome of your product, as well as being useful guides for shaping, developing and framing your arguments.

We have already argued that **writing an argument is a process**, and that many revisions are usually required to ensure that you produce an impressive product. As a result, ensuring that the content and form of your argument meet with your particular requirements is strongly advised, before you start the final checks of language, syntax and grammar (which will be covered in Chapter 11). Detecting issues with your reasoning or the quality of your claims at this stage – rather than later – could save you considerably more time and energy than if you wait until you have tweaked your product for language too.

Graham Coghill's (2019) *Science or Not?* website is an excellent resource for guiding you towards a scientifically sound method of discourse. The following page has great examples of 'red flags' that you should watch out for in argumentation: https://scienceornot.net/science-red-flags/. Remember to consider the fact-checking elements we dealt with in Chapter 5 too!

As part of examining your argument's reasoning and form then, you should revisit the six elements of the Toulmin method, and use each of the points to check your argument. Make sure that you can identify all claims, and that each of these has all the necessary components (at least the basic three, but preferably all six). Once you have checked off these items, make sure that your argument meets the most suitable requirements of Aristotle's rhetorical strategies, and, in particular, focus on how you have used the GASCAP arguments.

 For some practical activities on how to do this check, visit the Learning Zone to complete a guided Toulmin model analysis activity (do the assessment Chapter 9 Toulmin model analysis activity as well as the student exercise LZ 9.1).

In addition to these checks, make sure that you have not fallen into any logical fallacy traps, and try to be honest about whether or not you have managed to examine your topic from as many viewpoints as necessary.

Remember that **critical reflection** is an incredibly valuable tool: if you cannot be honest about your work to yourself, you are not likely to pass Aristotle's credibility test either.

To try and achieve this, ask yourself if you are certain that all other readers would draw the same conclusion that you have as you are assessing each element of your argument. Approach your claims and evidence in a critical way, and from as many different angles as possible. In the *unlikely* case that this check results in the conclusion that all readers would end up agreeing with you, your argument is probably sound, if you have been thorough and honest with yourself.

In the much *more* likely scenario that some readers would not agree, or that there might be some differences in understanding, ensure that you refine the statements to remove any possible ambiguity or evidence of fallacious logic. Where necessary, find different evidence or different ways of expressing it to support your claims, and deal with any counterarguments in one of the three ways discussed earlier to reduce their effects on your argument. The more you revise, the more reliable and defensible your argument should be; so *before* you submit arguments for assessment, use the various strategies in Chapter 11, as well as the tools provided in this chapter.

9.5 Conclusion

This chapter examined the concept and nature of complex problems, and demonstrated some key tools to use in developing logically defensible arguments, including both the Toulmin method and Aristotle's three rhetorical strategies. Importantly, both of these tools (in addition to the GASCAP argument strategies) were identified as useful items to evaluate arguments, if they are used in a critically reflective way.

References

'Abdu'l-Bahá. 2020. 'Tablets of the Divine Plan, 14: Tablet to the Baha'ís of the United States and Canada, 1917'. *The Baha'í Faith*. Retrieved 09/01/2020, available at <https://www.bahai.org/library/authoritative-texts/abdul-baha/tablets-divine-plan/15#459274152>.

Alderman, L. 2015. 'Recounting a bustling office at Charlie Hebdo, then a "vision of horror"'. *The New York Times*, 8 January 2015. Retrieved 30/08/2019, available at <http://www.nytimes.com/2015/01/09/world/europe/survivors-retrace-a-scene-of-horror-at-charlie-hebdo.html?_r=0>

Aristotle, 350 BCE. *Rhetoric, Book III*, translated by W. R. Roberts. Retrieved 30/08/2019, available at <http://classics.mit.edu/Aristotle/rhetoric.3.iii.html>

Bernstein, D. 1990. 'A problem solving interpretation of argument analysis'. *Informal Logic*, 12(2): 79-85.

Coghill, G. 2019. 'Science red flags'. *Science or not*?. Retrieved 30/08/2019, available at <https://scienceornot.net/science-red-flags/>

Dickens, C. 1981. *A tale of two cities*. Classic edition. New York, NY: Bantam Books.

Hornby, A. S. 2010. *Oxford advanced learner's dictionary of current English*. 8th edition. Oxford: Oxford University Press.

Larson, A. A., Britt, M. A. and Kurby, C. A. 2009. 'Improving students' evaluation of informal arguments'. *The Journal of Experimental Education*, 77(4): 339-366.

Ross, K. L. 2019. 'Some moral dilemmas'. *The proceedings of the Friesian School, fourth series*. Retrieved 30/08/2019, available at <http://www.friesian.com/valley/dilemmas.htm>

Shakespeare, W. 1998. *Hamlet*. Updated edition. New York, NY: Simon & Schuster

Spector, J. M. and Park, S. W. 2012. 'Argumentation, critical reasoning, and problem solving'. In S. B. Fee and B. R. Belland (Eds.). *The role of criticism in understanding problem solving: Honoring the work of John C. Belland*: 13-33 New York, NY: Springer-Verlag.

Toulmin, S. E. 2003. *The uses of argument*. Cambridge: Cambridge University Press.

Werry, C. 2019. 'Argument Framing – Chris Werry in The uses of argument'. *Course Hero*. Retrieved 30/08/2019, available at <https://www.coursehero.com/file/11166572/argumentframing/>

Wilde, O. 1997. *The critic as an artist*. Green Integer Series No. 3. Los Angeles, CA: Green Integer.

CHAPTER 10

Working with graphic devices

'... figures will not lie [yet i]t is equally true that liars will figure...'
– H. G. Wadlin (1890:283 1(6))

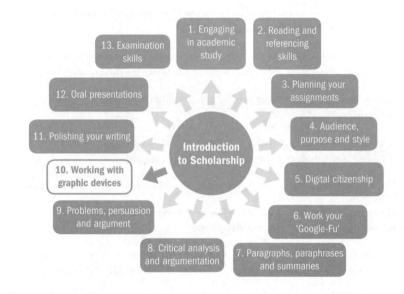

- 13. Examination skills
- 1. Engaging in academic study
- 2. Reading and referencing skills
- 12. Oral presentations
- 3. Planning your assignments
- 11. Polishing your writing
- **Introduction to Scholarship**
- 4. Audience, purpose and style
- **10. Working with graphic devices**
- 5. Digital citizenship
- 9. Problems, persuasion and argument
- 6. Work your 'Google-Fu'
- 8. Critical analysis and argumentation
- 7. Paragraphs, paraphrases and summaries

OBJECTIVES

At the end of this chapter, you should be able to:
- demonstrate familiarity with common graphic devices;
- interpret and describe given graphic devices;
- recognise and avoid bias in graphic devices;
- use given data to generate simple digital graphic devices;
- evaluate graphic devices for accuracy and user-friendliness.

Having been forced to take mathematics at school despite being abysmal at the subject, Alex had felt enormously relieved when he had passed matric and enrolled for a maths-free Social Sciences degree. For his upcoming assignment on xenophobia in South Africa though, he has noticed that almost all of his sources use tables, charts and graphs to present their data. He feels like he is doing high-school maths all over again.

Alex can *sort of* make his way around the tables, but some of the graphs and charts in the literature are horrifically complex. The worst part is that his lecturer has told his class that they need to include their *own* digitally created tables and graphs in their assignments, and he has no idea which ones are suitable, or how to actually create them.

When his friend Rashid suggests that Alex uses Microsoft Excel or Word to generate graphs, he feels relieved and hopeful that the problem can be resolved by his trusty computer. But after ten minutes of trying to enter numbers and headings, Alex tells himself that he may just have to accept that he *still* doesn't understand what these figures do, what information he needs, or even if the computer *can* generate what he *thinks* he needs. How is he supposed to know which *type* of graph to use, let alone where to put his information?

10.1 Introduction

Alex's struggle is a common one. Although graphs, charts and tables are included in texts to try to simplify and summarise findings, many readers find it difficult to interpret them. Something that made perfect sense to the creator of the graphic (who had the benefit of spending a great deal of time with their numbers and research) may be incomprehensible to the reader. One of the best remedies for avoiding confusion is to ensure that graphics are always contextualised by explanatory paragraphs, but this advice is unfortunately frequently ignored, with the result that readers are often left confused by these complex graphics that have been placed in an apparently random way.

However, what's worse is that we sometimes simply accept what charts, graphs and tables say – without critically analysing them – because they often present hard figures and numbers. We might be trained to identify bias and fallacy in writing, but we are usually not made aware of similar tricks or problems in graphics. We sometimes end up accepting misrepresented data, and then forming incorrect conclusions on a subject, because we do not consider that these devices may distort reality horribly.

This chapter will introduce you to the graphs, tables and figures that you are most likely to come across in the course of your undergraduate degree, and will guide you through an approach to describing and interpreting them. As an important element of this chapter, we will examine how graphs and figures are sometimes used to misrepresent data, and so help you to evaluate the 'honesty' of these kinds of graphic devices. The final section of this chapter will help you to generate your own graphic devices for assignments using simple Microsoft Word and Microsoft Excel tools. Because of the practical nature of this chapter, the most important part of it lies in the many hands-on activities that you will find for it on the Learning Zone. It is therefore vital that you try as many of these as possible in order to practise the skills of successfully interpreting and reporting on graphic devices.

10.2 Common graphic devices

You have probably needed to interpret graphs, tables and figures during your school career. To some degree then, you should already have a rough mental model (see Chapter 9 for more on this) of how to interpret these devices. However, it is unlikely that you ever took part in a specific school lesson on working with these, so you may find this section useful for getting to know some of the most common types and uses of graphic devices in academic literature, and how to analyse them. Before we begin, it is useful to recognise the main purposes of graphic devices.

10.2.1 Functions of graphic devices

In an academic context, graphic devices such as graphs, charts and tables can be used to:

- show groups of numbers or percentages that might otherwise be difficult or tedious to explain in a paragraph;
- demonstrate relationships between different things that would be more difficult to explain in words;
- describe data;
- provide a quick visual summary of information.

From these uses, you can see that the primary purpose of these devices is to summarise and simplify information, and to show relationships between that information in a way that is visually impactful.

Remember that graphic devices are an *additional* tool in academic writing; for them to be effective, they should always be contextualised by an introduction, and/or explained further by explanatory paragraphs.

Although there are a number of different types of graphic devices (from simple images to highly complex mathematical graphs), for the purposes of this chapter we will examine three main kinds of graphic devices: tables, graphs, and charts.

10.2.2 Tables

Tables are likely to be the graphic device that you are most familiar with, since they are commonly used in high school curricula to organise data, and feature in many other contexts too (for example, think of an Eskom load-shedding schedule, or even a television guide format).

A **table** is a grouping of information from different categories arranged into **rows** (which run horizontally) and **columns** (which run vertically), and is a highly useful way to organise information, especially if you are going to convert any of it to a graph later. Tables are usually vertically laid out, but they can also be horizontally formatted, depending on the categories you are working with.

To investigate the different elements of a table, let us take a look at Table 10.1 below, which tabulates the possible combinations of blood types that may result from parents with the same or different blood types. The various components of tables in general are explained in the section that follows by referring to the particular sections of Table 10.1.

Table 10.1. Possible human blood type combinations

Mother's blood type	Father's blood type		
	A	B	O
A	AA (Type A blood)	AB (Type AB blood)	AO (Type A blood)
B	AB (Type AB blood)	BB (Type B blood)	BO (Type B blood)
O	AO (Type A blood)	BO (Type B blood)	OO (Type O blood)

- **Table number**
  - > A table is numbered either according to the chapter in which it appears (e.g. the tables in Chapter 2 will start at 2.1 and proceed to 2.2, 2.3 etc.), *or* continuously by actual number of appearance (e.g. the first table in a book will be Table 1, regardless of which chapter it appears in), depending on where they are being used.
  - > Our table is labelled *Table 10.1*.
  - > Note that tables should always be indicated as *Table X* in academic texts, but graphs and charts are both indicated as *Figure X* (if you have a graph followed by a chart, the graph might be labelled Figure 1, and the chart labelled Figure 2).
- **A table title (next to the number)**
  - > The title of a table should be as descriptive as possible, without waffling.
  - > Our Table 10.1 is titled *Possible human blood type combinations* (even if you are do not know much about biology, it is clear that the table relates to possible blood combinations and that it is restricted to humans, so it has done its job in describing what the table is about).
  - > If you have sourced your table from somewhere else (i.e. not created it yourself), you need to cite your source underneath the table as shown here, and provide a reference list item for it.

- **Column heading**
  - > Each column must clearly indicate what data it describes, via a heading.
  - > In our table, there are two main column headings: *Mother's blood type* and *Father's blood type*.
  - > Column headings apply to all sub-heading items underneath them, so in this case, while all of *Mother's blood type* options (A, B and O) are included in column 1, the *Father's blood type* options of A, B and O are spread over three columns (columns 2, 3 and 4) so that we can compare them when matched to the mother's blood types. Columns 2, 3 and 4 therefore all refer to the possible types of *Father's blood*.
  - > In Microsoft Excel, columns are given letter numbers (i.e. A–Z, then AA, AB etc.) to identify them.
- **Sub-headings**
  - > Each column and row must have a clear heading that describes what data is included; if necessary, it should indicate the unit of measurement too (tables often present percentages; in this case, you would show a % in brackets next to the sub-heading).
  - > In our table, the row headings and column sub-headings are the same (i.e. *A*, *B* and *O*), which allows us to determine different blood-type combinations easily.
  - > In Microsoft Excel, rows are labelled with numbers from 1 at the top downwards.

 To practise reading and interpreting different tables, visit and complete the Learning Zone student activity (LZ 10.1.1).

- **Cells**
  - > In a table, a cell is simply the block that is created at the meeting of the column and the row in which you will usually find data.
  - > In our table, there are a number of cells, most of which include the different combinations of blood types possible.
  - > In Microsoft Excel, you can refer to a cell by using its column letter (e.g. B) and row number (e.g. 7) – for example, cell B7; this cell will be in the second column (after column A), seven rows down from the top.

For reading this table, and most others, we need to read the columns and rows, and then find the relevant cell with the information we are looking for. In this way, tables are excellent tools for presenting data, and for helping us to compare data easily. For example, if we look at Table 10.1, we can determine that a baby whose mother has Type A blood, and whose father has Type O blood, will have Type A blood (see highlighted cell).

Although tables can help to simplify information, extremely detailed tables are not suitable for lay or non-academic audiences, so if you have a lay audience, you should try to restrict the volume of your data as far as possible. It is also useful to use Microsoft Excel or similar software products if you do wish to generate tables, as these reflect

A **lay** audience or **layperson** refers to a group or individual who is unfamiliar with or who does not have specialised knowledge of a particular subject, so that using jargon or technical language when communicating with them is likely to result in confusion.

the relevant columns and row numbers, and can generate a variety of other graphics from their data sheets. We will deal with this in more detail in Section 10.4.

10.2.3 Graphs

Although graphs are frequently used in academic contexts, they are sometimes tricky to understand – especially if you are dealing with statistical distributions or financial graphs. Fortunately for these more complicated graphs, your lecturer and subject textbooks will usually guide you through reading and interpreting them. However, it is important that you have a clear understanding of at least the more basic elements of graphs in order to ensure that you can navigate your way around the more mind-bending ones.

Graphs come in many different types and sizes, with many different functions. For example, if you click on the 'Chart' function in a Microsoft Excel or Word document, you will notice that Microsoft offers line graphs, bar graphs, histograms, area graphs, scatter graphs, and several others. Line and bar graphs are by far the most common types of graph that you will deal with. Bar graphs are suitable for showing separate sets of information, while line graphs are better for showing continuous relationships. Let's take a look at the most common components of graphs by examining the (fictional) Figure 10.1 as an example.

 Although you are unlikely to come across more complex line graphs in your first year of study, if you would like to find out more about them, please visit the student exercises for Chapter 10 on the Learning Zone for examples and descriptions of them (in section LZ 10.3 External activity and resource links).

Figure 10.1. Comparison of support for illegal strike action by Union X and Union Y teachers between 2016 and 2019

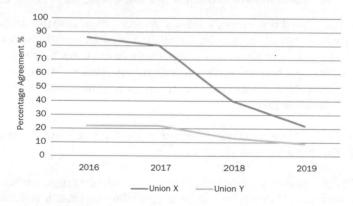

* **Figure number**
  > As with tables, each graph should have a specific figure number, dependent on which chapter or section it is included in, or how many other graphs or charts come before it.
  > We can label our graph *Figure 10.1* since it is the first graph of this particular chapter.

- **Figure title or caption** (next to the figure number)
  - > As with tables, the title should be as descriptive as possible, without providing unnecessary information.
  - > For our graph, we have the long but informative caption: *Comparison of support for illegal strike action by Union X and Union Y teachers between 2016 and 2019.*
- **Vertical axis**
  - > For our graph, the vertical (upright) Y-axis measures *percentage agreement*, but could measure a number of different units, including individual items, sales in rands or another currency, weight, or numbers of incidences etc.
  - > It is important that axis labels are present as these help you to understand the data – in this case, the label is 'Percentage Agreement %' so we know that we are dealing with percentages and not individuals.
- **Horizontal axis**
  - > For our graph, the horizontal (flat) X-axis measures the *years* in which the same question is measured, and reports on these (2016–2019).
  - > In a line graph of this nature, the data usually reflects a trend in one direction or another.
  - > Again, it is vital that clear labels are provided for each point on the graph so that readers can understand the information presented.
- **The key or legend**
  - > The key, which explains what the lines on the graph refer to, can be provided in a separate little box on the side of the graph or below it, depending on the style of the graph.
  - > For this graph, we have just two lines, and so the key indicates that one line refers to Union X, while the other refers to Union Y.

If you analysed this graph, it should be pretty clear that the support of teachers in both unions for illegal strikes has fallen dramatically between 2016 and 2019. This is especially so in Union X, where the majority of teachers originally supported illegal strikes. The difference is less notable in Union Y, but the trend away from support for these strikes is still evident.

 For more practice with reading and working with line graphs, visit the Learning Zone for student activity LZ 10.1.2.

Bar graphs are also popular figures and you will certainly come across and be required to interpret and create these. They are particularly useful for comparing different amounts, and are most frequently presented vertically, although it is also acceptable to present them horizontally if this suits the data. You are likely to encounter single, multiple, or perhaps even cumulative, bar graphs during the course of your studies. In order to demonstrate a bar graph, let us look at another hypothetical example: Figure 10.2 that follows.

Figure 10.2. Comparative sales figures for Africa 2017–2019: QT Company

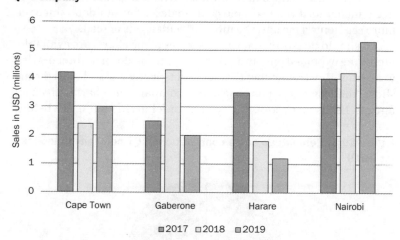

If you examine this graph, you will notice that each branch of QT Company has three bars, each of which represents a different year (see the key below the X-axis). The larger gap between each branch's cluster of results is useful for comparing how the different branches performed in terms of sales, which are reported in millions of US dollars. For example, in 2019, Cape Town's branch made sales of $3 million, up from 2018, but down from 2017.

Let's take a closer look at what this figure tells us. For example, it is clear that the Nairobi branch performed better overall than any other branch did. Although both Cape Town and Gaborone each had one good year, Nairobi had the most successful year of all in 2019, and its previous two years were also consistently good. Would you be able to tell which branch was in trouble by looking at this graph? How would you support your conclusion? How do you think managers of this business could use the data?

 Bar graphs are frequently used in academia so please make sure that you are familiar with this form of graph. Use the student activity on the Learning Zone (in section LZ 10.3 External activity and resource links) to help you practise.

10.2.4 Charts

The pie chart, or area graph, is another type of graph with which you are no doubt also familiar. Pie charts usually show the percentage split of an overall amount, and resemble a pie with slices of different sizes. Pie charts are usually easy to interpret, and are useful and appropriate when working with a lay audience. However, they do not provide very accurate measurements, and so should probably be avoided for more exacting subjects.

As with other graphs, all pie charts should still be labelled with a figure number and caption that clarifies its subject; if the pie slices are not directly labelled, then a key must accompany it. The largest

If you are interested in more advanced types of graphic devices, Heer, Bostock and Ogievetsky (2012) offer an interesting 'Tour through the visu-alization zoo' as part of a Stanford University presentation; it's available at http:// homes.cs.washington .edu/~jheer/files/ zoo/. Their article features a number of 'clickable' charts which you can interact with – if you don't 'get' all of these charts, don't feel intimidated, since a number of advanced devices are included, which you are not likely to use in your under-graduate programme.

portion of the pie is usually situated at the 12 'o clock position working clockwise towards the smallest value. Avoid using more than four or five data points if you use a pie chart to describe your data, since the more segments a pie has, the more difficult it gets to read.

You should be aware that various styles of pie charts are available, including exploded pies and 3D versions. If you do encounter or wish to use these, just remember that some perspectives may distort the pie slices; in these cases, it is wisest to accompany the chart with exact figures. Let us take a look at a 'typical' pie chart in Figure 10.3 below.

Figure 10.3. Percentage sales contribution of QT company branches for 2019

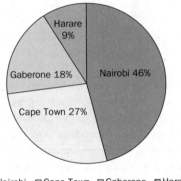

◼ Nairobi ◻ Cape Town ◻ Gaberone ◼ Harare

 Please visit the Learning Zone, which offers pie charts for you to practise your interpretation skills on (see student activity LZ 10.1.3).

If you take a quick look at Figure 10.3, it is easy to see that Nairobi contributed the most towards QT Company's 2019 sales figures, followed by Cape Town. There may be a number of reasons that Gaborone and Harare are not contributing to the same degree (e.g. they may have much smaller branches), but if all the branches were equal in size and potential capacity, then someone looking at this chart might be concerned about the comparatively poor contributions of these branches. Alternatively, they may be delighted by Nairobi's performance.

You should now have a pretty good idea of how to approach graphs and tables, and what to look for when you are presented with them, even if you do not yet feel like an expert in reading and interpreting them. This next section will help you to identify misleading elements of graphs and figures: it is vital to helping you develop your critical reading skills.

10.3 Detecting and avoiding bias in graphic devices

If you have ever seen a Fox News statistical graph, you will probably already be familiar with misleading graphs.

For a great clip on some examples of this, visit David Pakman's YouTube clip entitled 'Exposed: How Fox News lies with statistics', at https://www.youtube.com/watch?v=w7EvBxRYNME

Although there are a number of different ways in which a graph might mislead an audience, some of the main tricks (or mistakes, if you give the creator the benefit of the doubt) in presenting graphs are given in this section.

10.3.1 Not starting the Y-axis at 0

In fictional Figure 10.4a, it appears as though Medicine A has a far greater efficacy than Medicine B. In fact, Medicine B looks like a pretty awful choice. When you examine the Y-axis though, you might notice that it begins at 80% (already a very high efficacy) and ends at 90%. More importantly, Medicine B is actually just 8% different from Medicine A in terms of efficacy – by starting the Y-axis at a greater value than 0 (and in this case, ending it at 90%), we have grossly distorted the results.

Figure 10.4a. Distorted Y-axis graph

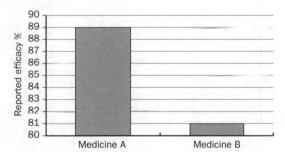

If the Y-axis were corrected by starting at 0 and ending at 100%, the two medicines would both look impressive in terms of efficacy, but Medicine A will still reflect a slightly higher rate. Let's take a look at fictional Figure 10.4b, which corrects this distortion.

Figure 10.4b. Corrected Y-axis graph

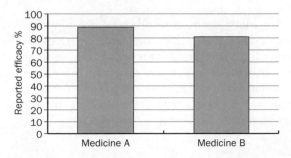

Can you see that with the axis restored to its honest proportions, the reported differences between Medicine A and B appear to reduce substantially? Keep this in mind whenever you are reading or viewing graphs – make sure that the Y-axis begins at 0.

10.3.2 Using different interval spaces on any axis

Changing the starting point of the Y-axis to something other than 0 can distort data, but using an uneven interval on this axis can grossly misrepresent data. Let's take a look at fictional Figure 10.5a below, which compares the prices of tickets for different sports in rand values.

Figure 10.5a. Comparative pricing of South African sports tickets

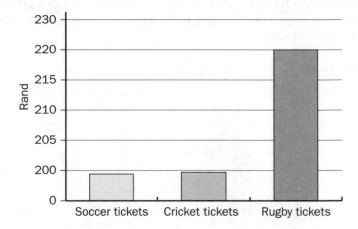

From this graph, rugby tickets appear to be almost three times the amount of cricket and soccer tickets! However, if you look at the Y-axis, you will notice that the first unit of measure (in rands) is a massive interval of 200, while the rest of the units are all just five intervals apart. This means that amounts over R200 will appear to have a hugely disproportionate relationship to the tickets that are under R200.

When you actually examine the prices by looking at the Y-axis carefully, rugby tickets are certainly more expensive than both cricket or soccer tickets, but not as much as they appear in Figure 10.5a.

To correct this, you need to use regular intervals on the Y-axis (and remember to begin the axis at 0), as in Figure 10.5b below.

This version of the graph reflects the data much more accurately than the original.

Figure 10.5b. Corrected Y-axis interval graph

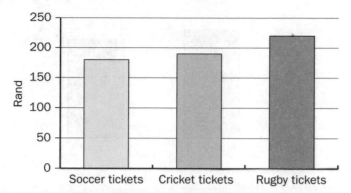

10.3.3 Incomplete data

Many graphs fall short because, either deliberately, or mistakenly, they do not include all the information that we need to determine whether the information is credible or not. This is done in a number of ways:

- Completely omitting a Y-axis is common; in these instances, we are often offered random bars that favourably compare what the author is trying to sell us – but we have no idea what the bars represent.
- Failing to include titles, captions, keys or legends, axis labels or units of measurement is also very common – if any of these are missing, try to find the information in the explanatory sections immediately before or after the graph, or conclude that the information might be untrustworthy (more likely). For example, would Figure 10.4b make any sense at all if its Y-axis label 'Reported Efficacy %' was absent? We would then simply have two medicines being compared, but we would have no idea how they were being compared; the graph thus becomes useless.
- Important contextualising information is also often excluded – if you were paying attention in our earlier example of Figure 10.4b, you might have noticed that we measured the reported 'efficacy' of two different medicines. We very favourably reported that Medicine A had an almost 90% efficacy, making our product look particularly useful. But what does this efficacy refer to? A study of three siblings? A group of 100 volunteers? A controlled experiment sample? This is important information – even if the graph is formatted correctly, we should not accept its data as the truth without critical engagement! Providing clear context is ethical. This can be done in the text that describes the graph, but should ideally be included on the graph itself, as seen in Figure 10.4c below.

Figure 10.4c. Corrected Y-axis interval graph

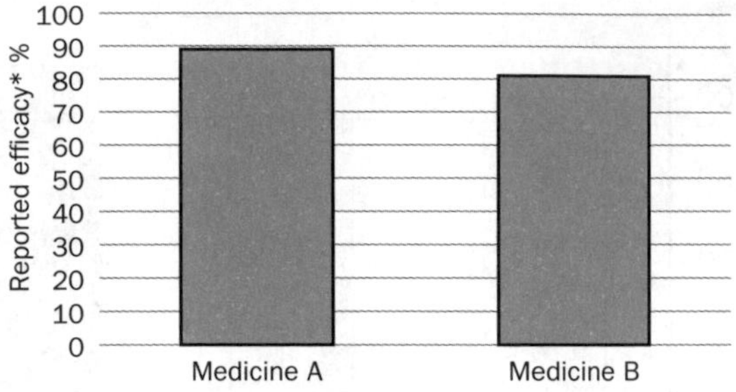

* Based on a controlled study of 1 000 patients

10.3.4 1 + 2 = 8 data

A **percentage** is a ratio or number expressed as a fraction, or part, of 100. For example, if you got 35 out of 50 for a test, this would mean that you got 70% since 50 can go into 100 twice, and you would then need to multiply 35 by 2, which is equal to 70.

The term **mutually exclusive** means that two things cannot both be true at the same time. For example, a light switch can be on or off, but not both on and off at the same time. On and off are therefore mutually exclusive categories.

Using numbers that simply do not add up is the final problem that we find in graphs and charts. For example, if you watch Fox News (truly, we are not picking on them, but their graphs do beg for amused analysis), you might find pie charts that add up to more than 100%. Since the whole point of a pie chart is to demonstrate how elements are apportioned in percentages, this is a significant problem. If you are using a pie chart, you must convert item quantities to percentages – regardless of how many items you are measuring.

A related problem occurs when items that are not mutually exclusive are used in a pie chart. For example, if you were to measure the age of rhinos in a rehabilitation camp and divided their ages into three categories (0-2 years; 2-4 years and 4+ years), rhinos who are 2 years old fit into the first and second categories, while rhinos who are 4 fit into the second and third categories. If we then attempted to plot our rhinos' age on a pie chart, we might come to a very confusing total in excess of 100% because we may have recorded some rhinos more than once. To avoid this, we should instead choose **mutually exclusive** categories like 0-2 years; 3-4 years; and 5+ years. Our pie chart would then add up to a more sensible 100% as we can suitably assign each rhino just once, into a single category.

For this reason, it is essential that graphs are careful about what data items they claim to represent, and that any areas of confusion or distortion are noted in your descriptions – better yet, if you find a problematic figure in your sources, rather search for a different source. While mistakes do happen, it is safer to stick with a source whose author has clearly paid attention and checked their work!

10.4 Generating graphic devices using software

Okay, so you now have a pretty good idea about what types of graphs you might need to deal with, and how to check that the information you have sourced is on the level. However, you still need to know how to generate your own graphs. In this, you will be happy to hear that Microsoft Excel and Word make generating graphs, tables and charts a lot simpler than it used to be. This section will help familiarise you with the appropriate part of the software that will create your graphs for you, and use an example to demonstrate how it works. For the purposes of this section, we will develop a chart comparing the performance of three different groups of first-year students in an academic literacy programme at University X and use Excel to produce a graph for us.

Step 1

- Draw up a table with the relevant data that you need to include.
- You can either paste this data into Excel or create a table using Word.
- For this example, we will create a table in Word with five columns, and four rows (Click the 'Insert' tab, click on table, and move your mouse over the blocks until you have five columns and four rows);
- Add in headings and the specific data as indicated below:

Group	0–24%	25–49%	50–74%	75–100%
Group 1	1	0	12	2
Group 2	0	2	13	3
Group 3	3	3	10	0

- As you can see, the numbers entered in each cell represent the number of students in each group who achieved within the given percentage.
- Please note that you may have arranged your data in the opposite way (i.e. groups along the x-axis and percentages along the Y-axis). Microsoft Excel allows you to switch rows or columns so this should not make too much of a difference.

Step 2

- Now for the truly miraculous part of the software!
- Copy the table data that you created in Step 1.
- Paste it using the 'Match destination formatting' into any cell on Excel.
- Now select all the information, and click the 'Insert' tab (do not just right-click on the selected information).
- You now have the option of selecting from pie charts, scattergrams, line graphs, and a number of more complex options. For your purposes, use the special Excel button called 'Recommended Charts', which evaluates your information and suggests the graph that would be most suitable for your purposes.
- Click on 'Recommended Charts'. You will notice that the first item offered is a 'clustered column'. As we are trying to compare the results of three different groups across different percentage

divisions, this is a suitable chart for our purposes, although it is not the only one we could have used. Click 'ok' and you will see that your data has been converted into a chart.

Step 3

- The problem with the resulting graph is that the items are grouped by percentage, and this is a bit awkward to draw comparisons from. Since we would like to view each group's performance in relation to the others', we would rather the graph used this as the organising structure. Click the 'Switch Row/Column' button, and the image that pops up should look similar to Figure 10.6a below (although yours might be in colour). If you are working in a Word document, you can now copy and paste your chart into your document from Excel.

Figure 10.6a. Comparison across academic literacy groups for 2020

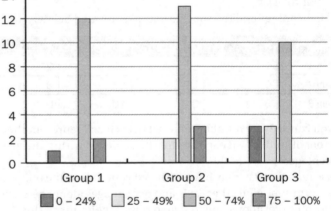

- At the moment, we do not know what the student numbers on the Y-axis refer to since there is no label on that axis, although there are appropriate student numbers on it. To add in a label, click on the right side of the graph in Excel, and click the big + sign, then tick the 'Axis title' block – once this appears, type in 'Student numbers'.
- Ensure that your figure has a complete caption and number.
- Our graph also still says 'Chart title' at the top – since we have a different format caption at the bottom of our graph, we can delete the chart title on the graph itself.
- If you wanted a more detailed table included with your graph, you could click the + sign again, and tick the 'data table' option. This will include the information from your original table neatly under the graph itself. This is sometimes useful, but be careful of cluttering your figures – the point is to help explain and summarise, not to overburden your reader with information.

- If you completed all of the above instructions, your figure would end up looking something like this:

Figure 10.6b. Comparison across academic literacy groups for 2020

	Group 1	Group 2	Group 3
■ 0 – 24%	1	0	3
□ 25 – 49%	0	2	3
▨ 50 – 74%	12	13	10
■ 75 – 100%	2	3	0

■ 0 – 24%   □ 25 – 49%   ▨ 50 – 74%   ■ 75 – 100%

Now that our graph is complete, it is easy to see the particular trends for each category. For example, we could easily say which group appears to be doing more poorly than the other two (Group 3, with just 16 students, had six students who failed the course, in comparison with Groups 1 and 2, who together have a total of three students who failed. Group 3 also failed to produce any students with distinction marks, but both Groups 1 and 2 did so. Finally, Group 3's performance in the 50–74% category was slightly below that of Group 1 and 2, so, overall, the group definitely did more poorly.)

Although you can conclude from the graph that Group 3 did more poorly than the other two groups, note that you will not be able to explain from the graph *why* Group 3 did less well. This is an important point with regards to working with and interpreting graphic devices: the data and the images provide you with information, but they do not necessarily explain the causes behind the data. As a result, you should be particularly careful of deducing conclusions that are simply ideas. For example, in the above example, we might wish to argue that Group 3's lecturer was simply not as good as the lecturer of Groups 1 and 2 – or we might argue that the students did not attend classes. However, we cannot draw any real conclusions without having some further evidence to back up statements like these.

For some detailed activity-based examples of different kinds of charts that you can generate, please complete the student activities in section LZ 10.2 and 10.3 on the Learning Zone.

10.5 Evaluating graphic devices

You should now have an idea of some of the most frequently used graphic devices, some of the ways in which these are used to misrepresent data, and also an idea of how to generate your own charts using digital software. Here is a final checklist that you can use both to check graphic devices and to ensure that your graphic is good-to-go.

A humorous (but really helpful) look at generating graphs is available on the HackerSpace website, at http://hackerspace .lifehacker.com/ 5-rules-for-making-graphs-1605706367 under the title '5 rules for making graphs'. Some of these overlap with the items below, but the short article is funny and very useful, so you may enjoy it anyway.

- **Check – and double-check – your data**
  - > Before you ask Microsoft to perform any miraculous creations, ensure that you have checked, and double-checked the data that you are entering – if your information is wrong, your graph or chart will *always* be wrong, regardless of how much time you spend making it look credible.
- **Provide any necessary explanation, key or legend**
  - > Remember that most of your printed work will be in black and white, so be careful when you use colourful graphs – they look great on your computer, but if the colours are converted to greyscale without you checking them, they may render your graph nonsensical.
- **Label your axes**
  - > We have mentioned this before, but remember how important it is to ensure that your reader understands the *context* of your data, and that your Y-axis starts at 0, and increases at a regular index rate.
- **Include units of measurement**
  - > These should usually appear in the Y-axis label, but might appear elsewhere, including on the graph bars themselves – just ensure that your reader knows whether they are reading about money, percentages, individual units, or something completely different.
- **Do the maths**
  - > Yes, yes, we know it is hard sometimes, but for simple things like pie charts, either let Microsoft do the calculations and present a pie chart, or, if you have done the calculations yourself, ensure that the numbers do, in fact add up to 100% – no more, no less.
- **Reference your source**
  - > As with all your work, if you have used another author's ideas or work, ensure that you indicate this under the figure or table as demonstrated earlier in this chapter, then include the full reference at the end of your assignment.
- **On location – context matters**
  - > Every graphic device that you create should consider your audience: some graphic device formats, like pie charts, are more suited to lay audiences than expert audiences, and vice versa. Depending on what you are designing your graphic for, you might also need more or less detail, colour, or size. Use a font size and font that is suitable for the context, and ensure that you use the device for what it was intended: to *convey meaning simply, honestly, and effectively.*

A number of student exercises that require you to interpret, describe, and evaluate different graphic devices are included as revision activities for this chapter on the Learning Zone (including an online assessment and a video). Please try to complete as many of these as possible – the more you practise, the more comfortable you will feel working with the data.

10.6 Conclusion

This highly practical chapter presented some of the most commonly encountered forms of graphic devices in the academic context, such as tables, charts and graphs. It also examined some of the most frequent ways in which these devices are made to deliberately distort otherwise reliable findings, and provided particular cautions on avoiding these pitfalls. The chapter also provided a practical step-by-step example of how to use Microsoft Excel to generate a bar graph, and offered many practical, guided, online activities on which to practise this skill. It concluded with a short checklist for ensuring that graphic devices are soundly put together and are reliable.

References

HackerSpace. 2014. '5 rules for making graphs'. *HackerSpace*, 15 July 2014. Retrieved 30/08/2019, available at <https://hackerspace.kinja.com/ 5-rules-for-making-graphs-1605706367>

Heer, J., Bostock, M. and Ogievetsky, V. 2012. 'A tour through the visualization zoo'. *Paul G. Allen School of computer science and engineering, University of Washington*. Retrieved 30/08/2019, available at <http://homes.cs.washington .edu/~jheer/files/zoo/>

Pakman, D. 2012. 'Exposed: How Fox News lies with statistics'. David Pakman Show channel on *YouTube*, 29 November 2012. Retrieved 30/08/2019, available at <https://www.youtube.com/watch?v=w7EvBxRYNME>

Wadlin, H. G. 1890. 'Seventh annual report of the Bureau of Labor and Industrial Statistics'. Boston, MA: Bureau of Labor and Industrial Statistics

CHAPTER 11

Polishing your writing

'I'm all for the scissors. I believe more in the scissors than I do in the pencil.'
– Truman Capote (1985:205)

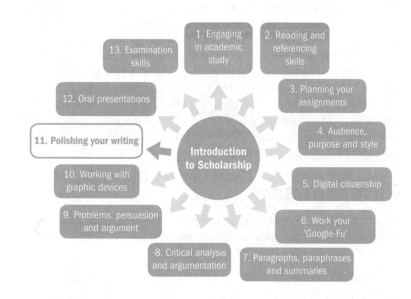

13. Examination skills

1. Engaging in academic study

2. Reading and referencing skills

12. Oral presentations

3. Planning your assignments

11. Polishing your writing

Introduction to Scholarship

4. Audience, purpose and style

10. Working with graphic devices

5. Digital citizenship

9. Problems, persuasion and argument

6. Work your 'Google-Fu'

8. Critical analysis and argumentation

7. Paragraphs, paraphrases and summaries

 OBJECTIVE

At the end of this chapter, you should be able to:

- follow given procedures for effectively revising, editing and proofreading written texts.

> ### IRL
>
> Last night, an exhausted Lerato collapsed onto her very welcoming bed after finally completing her Psychology essay. After spending many hours researching her topic then developing and revising her argument, she had spellchecked and printed her document, and felt more than ready to submit her assignment – a full two days ahead of schedule! It was with a great sense of weary accomplishment that she drifted off to sleep.
>
> However, on the bus to campus this morning, she made the fatal mistake of skimming through her assignment. What an awful idea! On the bus, Lerato noticed that some of her sentences and word choices looked, well ... weird, and that although she had what she felt was a good argument and credible support for it, the way that she had expressed some of her points did not always convey her argument clearly. Worse: some of her best points had been lost completely! When she started to read her essay more carefully, she also noticed a number of punctuation problems, missing full stops, and more than a few rambling sentences exemplifying what she knew her English lecturer would refer to as 'comma riot'. Feeling really annoyed with herself, she realised that she would have to delay submitting her assignment, and go back to the proverbial drawing board to revise and rewrite it.

11.1 Introduction

Lerato's experience is a common one – it may even depict the reason that many students choose *not* to review their completed assignments once they have printed or finalised them. But sticking your head in the sand like an ostrich is not an approach that is going to improve your marks. What Lerato thinks was a really bad idea, in the short-term, will actually almost certainly work in her favour in the long-term, when she is awarded better marks for her assignment.

The reality is that polish counts, and a good argument that is presented without errors and that reads well is much more likely to be seriously considered than a good, or even great argument that is riddled with problems. This is because first impressions count, as in many other areas in life. The assumption is that if you have taken the time and effort to write a grammatically and syntactically sound essay, then it is likely that you have also spent time and effort on building your argument. The opposite of this is also true: sloppy work that is in obvious need of editing makes a poor first impression, perhaps prompting a lecturer to dismiss the argument as weak too.

But what exactly happened in this case? Lerato did her research properly, she spent a good deal of time on building and revising her argument, and she even remembered to spellcheck her work before she printed it. Where did she go wrong?

Well, let's see. In Chapter 9, we explored building and analysing your argument claims according to the Toulmin method, Aristotle's three rhetorical strategies, as well as checking for logical fallacies and

> ### Point of interest
>
> Although the quality of your grammar and syntax is important, please don't think that spectacular academic language and sentence structure will let you get away with a poor argument. Although good language use can significantly improve the reception of your argument, markers are still likely to identify poor arguments, and should rightly call you out on them.

cognitive bias. To some degree, Lerato has checked for all these elements in her argument, so the problem should not be there. Lerato's main oversight is that she failed to revise, edit and proofread her final draft – as a result, her 'final product' is simply not polished enough to submit if she wants to get good marks.

Revise, edit *and* proofread, you say? All three of these steps *in addition to* the previous argument checks?

You might feel overwhelmed by needing to undertake more tasks than you were expecting, or perhaps you are confused about the difference between revising, editing and proofreading. Indeed, many people do confuse them, and so simply skip these processes. However, each of these tasks serves a different purpose: **revision** is an analysis of the global structure of the document and ensures that your content is suitable and logically ordered; **editing** analyses your syntax (sentence structure) to ensure that it is correct and supports your overall structure; and **proofreading** examines your document for grammatical and punctuation errors.

> **Perfunctory** means that something is done in a hasty, inattentive or slapdash manner. For example, 'Adam's perfunctory check of the parachute equipment terrified the trainees in the plane, and two refused to jump as a result – both were convinced that they would end up as bounce-rate statistics of the skydiving school.'

If you complete these steps and check your argument thoroughly, your submission has a much better chance of impressing your readers than if you ignore them, or go about them in a perfunctory way.

For this reason, this chapter will guide you through the steps of revising, editing and proofreading your writing and provide you with practical tips and strategies for ensuring that your writing is as error-free and impressive as it can possibly be.

11.2 First ... take a break

Wait ... what? Take a break? Yes, go have a KitKat.

As you can see from Lerato's experience, one of the main reasons we miss mistakes when we proofread immediately after completing a written piece is that the ideas themselves are still fresh in our minds. Because we have formed and shaped the arguments in our words in the world of our *heads*, instead of seeing and absorbing what is on the *screen*, we tend to 'see' or mentally hear what what we *intended* to say. Simply put, we see what we *intended* to write, as opposed to what we really wrote.

For a practical example of this, did you notice the extra 'what' in the previous sentence? If you did, well done! But many readers will not have noticed it. Although we all know that the extra 'what' is wrong in that sentence, your mind was most likely trying to interpret the *meaning* of the sentence, and so simply ignored or glossed over less important words. This kind of proofreading mistake is very common when we check our own writing immediately after completing it, with the result that we miss or repeat words despite actually knowing how to write a sentence correctly.

To avoid this, put some distance, in the form of time, between yourself and your work. For example, when you have completed what you feel is your final draft, save and close your document, and leave it for an evening. Get some sleep. Once you come back to it, the sentences and ideas that were swimming around your head should have disappeared. This will allow you to really read what is on the screen with a sharp eye and fresh mind, and help you to engage fully with the revising, editing and proofreading processes.

11.3 Next step ... revise

In checking your argument using the methods outlined in Chapter 9, you will have already completed a substantial component of your revision. Once you have completed the final revision steps, shown in Figure 11.1, your paper should be thoroughly revised and ready to move to the editing phase.

 To practise your skills in revising final drafts, visit the Learning Zone for student activity LZ 11.1, where you will find a sample argument and its matching skeleton outline, ready for your keen eye and revision!

Don't worry if you find this step a little exasperating – it is certainly one of the trickier parts of the essay-writing process, and the many different approaches and problems you uncover may leave you feeling very doubtful about the quality of your essay. For example, you may move paragraphs and sentences in your piece around, completely reworking the sequence, and then end up right back where you started when you decide to go back to your original sequence.

This is normal, and not a waste of your time! Considering alternative ways of doing things often helps you to find a better way of expressing your argument, or making it more persuasive, even if you do not always use the alternatives in the end.

Becoming more familiar with these tasks will also improve your ability to execute them. Given the number of assignments students are usually required to submit throughout their studies, you should be very adept at organising your writing by the time you graduate!

Meanwhile, you should be able to use this as a reliable guide for revision if you spend sufficient time and consideration on your skeleton outline (revisit Chapter 3 for guidance). When in doubt, let your skeleton dictate what goes where, and then just ensure that your transitions ease the flow of argument between the necessary sections of your argument.

11.4 Then ... edit

You now have an argument that you think you can be proud of: it presents credible, convincing evidence, and has a logical structure. What's next? **Editing**. This process, which is often confused with proofreading and revision, involves *careful questioning and analysis of sentences* that you have used to present your argument.

Point of interest

Some people prefer to revise their work on a printed version of their final draft and then capture the changes digitally, but the advantages of continuing to work on a digital copy of your document are that you can easily cut, copy, and paste sections and ideas, or move them around as necessary. It is also more environmentally considerate. Just remember to save new digital versions with the word 'revision' after your existing file name to be safe.

Figure 11.1. Steps in the revision process

Check purpose and main claim

> Check that your final draft actually serves the purpose described in your introduction.
> If either is out of sync with the other, revise your purpose or revise the content in your essay.
> Persuasive, argumentative, or comparative essays usually have a main claim – check that yours appears, and that you have supported that claim suitably in the paper.

Identify and check support for major claims

> Check that your main claim is supported by sub-claims, and that these are also suitably supported.
> Ensure that you have given sufficient supporting data in the form of reasoning or evidence AND that you have clearly explained how each item of evidence supports your claims.

Check against your skeleton

> Compare your draft to your skeleton outline.
> If there are differences, ensure that these are an improvement on the skeleton. If not, rework your piece to fit the original plan.
> Fill in any gaps in the argument as planned in the skeleton by providing extra detail, evidence or argument.
> Delete poor, or 'nice' but unnecessary, sections that do not add to your argument.
> Move paragraphs and sections of your essay around if you need to – just remember to tie all your work into the main claim and sub-claims in the most logical manner possible.

Identify and check transitions

> Identify all transition points between ideas in your draft (see Chapter 7 for more detail).
> Evaluate the effectiveness of these in terms of the writing 'flow' – your readers should easily follow the logic of your argument, and should be able to sense when your argument is about to turn, emphasise or develop a particular theme.
> Rework areas that feel 'jerky' or awkward by working on your transition techniques – revisit Chapter 7 to review examples of how to do this.

Reasoning strategies

Go through each paragraph and check the following, even if this requires a repeat revision:
> Ensure that each paragraph has a clear topic sentence, usually near the beginning.
> Ensure that the paragraph plays a clear role in the argument and that it is correctly situated in terms of main and sub-claims and support.
> Make sure that the particular argument structure and logic you have used makes sense for the type of essay you have chosen, and for the evidence you are presenting.

To do an effective edit of your work, you need to have a very good idea of the nature and structure of different sentences, which we covered in some detail in Chapter 7 itself, and in more detail in the Learning Zone sentence-focused activities for that chapter (LZ 7.2). For a quick refresher, examine the following nonsense sentence:

Blonky bombledums grunch sloggy hooplers.

Um ... what?

Okay, this sentence doesn't mean anything to you. Are you *sure* it doesn't though?

If I asked you who was 'grunching' in this sentence, are you sure you wouldn't be able to give me an answer?

What about if I asked you who was *being* 'grunched'?

Or what *type* of 'bombledums' I was talking about?

Could you identify what *type* of 'hooplers' were being discussed?

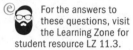

For the answers to these questions, visit the Learning Zone for student resource LZ 11.3.

If you think about these questions, you may realise that you were able to figure out the answers to them even though the sentence itself made little sense. This is because English sentences usually follow a particular format that we already have a mental model for (see Chapter 9 for more on this) as shown by the following:

subject + verb + object (optional)				
Blonky	bombledums	grunch	sloggy	hooplers.
(adjective)	(noun)	(verb)	(adjective)	(noun)

As a result, we can quite easily formulate a kind of understanding of what is happening in this sentence even though the words are nonsense, and their meaning unknown. This mental model of an English sentence (subject + verb + object [optional]) is very useful in the editing process; much of what you recognise as correct or problematic uses this model as the basis for your recognition. However, there are many different forms and shapes of sentences; your sentence types and style should always suit the audience you are writing for, as we pointed out in Chapter 4. So, please do brush up on your sentence 'anatomy' by re-reading 7.2.1 and visiting the Learning Zone Chapter 7 material before you edit your work.

Editing your work is both a reflective and practical exercise, so you need to ensure that you are completely honest with yourself when you do so. You will need to both question and analyse the sentences that you have used to convey your message, but you need not worry about grammar, spelling and punctuation errors now – instead you will identify and resolve these during proofreading.

To engage in this reflective editing process, you need to do three main things *for every sentence*:

- read it to check that it performs its function;
- analyse the sentences immediately before and after it to check the flow;
- and comprehensively edit it for length, purpose and simplicity.

Point of interest

We recommend completing the editing step in the digital copy of your document, just like the revision step. Some people prefer to edit on printouts of their work, but working digitally is simpler and more effective – it also saves trees! As with revision, remember to save your work in a new file with the word 'edit' after your existing file name just to be safe.

Each of these steps requires a little more explanation, so let's take a look at what you need to do in practice.

11.4.1 Sentence function

First, carefully read every sentence, asking yourself about its function in the paragraph in order to decide whether or not it does what it needs to. For example, if it is a topic sentence, does it clearly state your claim? Or if it's a supporting sentence, does it adequately provide evidence for a topic sentence? If an element of your sentence function is unclear, rework the sentence until it clearly achieves its purpose.

11.4.2 Before-and-after sentences

Once you have decided that your sentence performs its function, focus on the sentences immediately before and after it. Ask yourself if the *preceding* sentence provides a logical introduction or flow to your current sentence, and whether each *succeeding* sentence follows on logically from your current sentence.

Examining whether the transitions used are suitable may help you to determine this. If you sense an 'awkwardness', or realise that one sentence will create unfulfilled or different expectations for the reader, make sure that you rework, reorder, rephrase, or delete to avoid issues.

For example, if you look at the following three (hypothetical) sentences, you will notice that the first sentence creates an expectation that a particular type of information will follow, but the main sentence itself does not actually support this.

Despite clear evidence of the existence of xenophobic attitudes in South Africa, new evidence suggests that the majority of the country's citizens are not xenophobic. For example, in a study of 1 000 Johannesburg citizens (Rudolph, 2000), almost 80% of respondents agreed with the statement that foreign nationals 'steal' South African jobs. This figure is alarming, and suggests that much needs to be done to reduce the misperceptions regarding foreign nationals in South Africa.

The first sentence in this example creates two expectations. Firstly, it leads us to expect evidence that supports the idea that South Africans are actually *not* xenophobic. Secondly, it creates the expectation of *new* evidence, so we could reasonably expect something from the last year or two.

Does the second sentence deliver on these expectations?

No, the second sentence only supports the idea that South Africans *are* xenophobic, since almost 80% of respondents in our (fake) research findings think that foreign nationals 'steal' local jobs. This clearly does not support the statement in the first sentence and so we either need to change the first or the second sentence so that they agree with each other.

The second problem is that we have described the evidence that the majority of the country's citizens are not xenophobic as 'new'. However, the evidence presented in the second sentence has a publication date of 2000, making it quite dated, in addition to failing to agree with the main claim.

In this example, it is more logical to change the first sentence, because the final sentence appears to be warning readers that we need to reduce the misperceptions of foreign nationals. Changing the first sentence will change the expectation of the audience for the sentences that follow, and allow our evidence to support the claim. For example, we could change the first sentence to:

Despite evidence of anti-xenophobia campaigns in smaller segments of South Africa, some research findings seem to suggest that a much larger portion of the country's citizens hold xenophobic misperceptions.

If we replace the original first sentence with this new one, can you see that our evidence will make a lot more sense? This new structure works a lot better, since our argument, expressed in the final sentence, is that we need to do much to change the misperceptions relating to foreign nationals. It means there is no longer internal conflict between the ideas in each sentence.

We may also do well to add an extra source to support our first statement. Since our claim is quite a large one and runs the risk of being a generalisation, having more than one credible source will help our argument, and reduce the risk of not convincing our readers. We do not want to add a source to the existing middle sentence, because it clearly speaks about a particular study; instead, we could insert an extra sentence immediately after it, before the final sentence. If you do this, just make sure that you have adapted the concord and verb agreement between the sentences. In this example, you can see that we need to change a few elements of the final sentence to the plural.

Let's take a quick look at our new product then:

Despite evidence of anti-xenophobia campaigns in smaller segments of South Africa, some research findings seem to suggest that a much larger portion of the country's citizens hold xenophobic misperceptions. For example, in a study of 1 000 Johannesburg citizens (Rudolph, 2000), almost 80% of respondents agreed with the statement that foreign nationals 'steal South African jobs'. Another study by Mbeki and Morena (2013) arrived at a similar finding, with 76% of its respondent group expressing resentment towards foreign nationals for perceived encroachment on local jobs. These figures are alarming, and suggest that much needs to be done to reduce the misperceptions regarding foreign nationals in South Africa.

Right, are you happy with this edit? We checked that all three sentences work together – and ended up adding an additional one – and the argument now seems to flow more logically from one sentence to the next. What transitions did we use that allowed this to happen?

In the first sentence, the first word 'despite' indicates that a contrasting idea is coming up: this provides the first transition. The second part of the first sentence introduces 'some research findings' but does not immediately get to them. This leaves the reader with an expectation that we will provide details of these.

The second sentence dutifully complies with this expectation by leading in with 'for example', and so transitions quite simply from the previous sentence.

The third sentence also transitions neatly; by starting with the words 'another study', it clearly shows how it relates to the previous evidence in the previous sentence. The reader is led smoothly into the fourth and final sentence.

In the final sentence, the words 'these figures' tell us that the previous sentences have led up to this point, as the figures referred to are mentioned in the earlier sentences. A final observation on the (fictional) matter ends the sentence.

You can obviously tweak your transitions as you work with your sentences, so that they improve the flow of your argument, and deliver according to the expectations you have created. Revisit Table 7.1 in Chapter 7 for more specific examples of how you can work with transition elements in your writing.

Finally, for this step, remember to keep your skeleton outline handy so that your argument as a whole remains on track as you rework some of your sentences. Watch that you don't stray off topic – remember that the point is to ensure that all of your sentences ultimately contribute to your main claim and argument.

11.4.3 Evaluating every sentence

As a last step of editing, you need to actively evaluate every sentence in your essay. This might sound like a time-consuming and potentially super-boring activity, but the more you practise this, the easier and quicker it becomes – and the bonus at the end of it is that you will have a polished product.

During this stage, you should focus on three main aspects of your sentences:

Managing your sentence length

Academic writing can be complex, communicating profound, intricate ideas, but using long, tangled sentences may send your reader to sleep, or confuse him or her completely.

That does not mean that using hundreds of really short sentences is the solution – doing this could make your writing lack flow, making it difficult to see the connections between your ideas.

Instead, the answer lies somewhere in the middle: your sentence lengths should vary to achieve a balance. If you recall, in Chapter 7, we described topic, supporting and concluding sentences, with each of these serving a specific purpose. If your topic sentence is punchy and sharp, you could allow a slightly longer supporting sentence, then conclude the paragraph with another short or another slightly longer sentence.

Decide what you are trying to achieve with each sentence and paragraph, then choose whether to chop up a long sentence, or combine two shorter ones using a conjunction or transition. This may take some time, and you may well explore a number of alternative sentence lengths before you are satisfied. Don't skimp on this step. Make sure that you revise sentence length until you are satisfied that it is suitable for its place and function. You may still tweak your sentence lengths during proofreading, when you look at punctuation, but by the end of this stage, you should be satisfied with the variety of sentence types and lengths in your writing piece.

Making the grammatical core of your sentence the hero

In Chapter 7, we covered the three different sentence types that you will encounter in your paragraphs. In addition to ensuring that the sentences in each paragraph reflect those roles, you must also aim for the **grammatical core** of each sentence to be the most prominent part of it. The grammatical core is the subject + verb + object structure. Because academic writing often has layers of evidence, it is easy for the main structure of a sentence to be hidden by a structure that is too complex. This problem is especially common when we use the passive voice.

To avoid the grammatical core of a sentence being hidden, aim for your sentences to clearly show the real subject (or thing executing the verb), and the verb itself. Let us look at an example of the problem, and how to remedy it:

Analysis of the problem was done in terms of its nature and potential impacts.

You might understand this statement – it is not too difficult – but it is unclear who conducted the analysis because the real action or verb (the analysis) is in the form of a noun, and a rather poor verb (done) has taken its place.

To strengthen these two elements and make it clear who conducted the analysis, first identify the actor (your subject) and use the real action as a verb. Let's see how this could look in a less formal paper:

I analysed the problem in terms of its nature and potential impacts.

If you are working on a more formal paper and need to use the third person:

The author analysed the problem in terms of its nature and potential impacts.

In order to achieve this, we moved the real verb (*analysed*) as close as possible to the subject (*the author*), which is now actively identified. Having the verb near the start of the sentence helps the reader to process the meaning of the sentence more easily, and makes the grammatical core of the sentence more obvious.

Let's take a look at whether this works in the passive voice:

The problem in terms of its nature and potential impacts was analysed by the author.

Can you see that the version in the passive voice does not show its grammatical core as clearly as the previous version does? The reader has to remember a big chunk of information (*The problem in terms of its nature and potential impacts*) before he or she can get to the verb (*was analysed*) and the person who took the action (*the author*). The sentence is also longer. Both of these elements make the reader's task more complicated than it needs to be.

Aiming for representation

It is the 2020s but sadly, sexism, homophobia, racism, xenophobia and other prejudices are still alive and well. We should know better by now, but, judging from news reports, we seem to prefer to cling stubbornly to old ways of thinking. The fact is that much of our worldview is built on the structures and language around us; although we have improved, we are still surrounded by traditional patriarchal and heteronormative values. For example, why do adverts almost always show a woman doing the laundry? Why is mommy+daddy+2.0 children considered the 'normal' family? Why are most of my books written by grey-bearded men? Why does my boss ignore my ideas until my male colleague suggests them? Why do I still need to 'come out of the closet'? Why are 'flesh'-coloured plasters still white? ...

Clearly, we could go on all day here. Sadly, there is no quick fix to social discrimination, as we pointed out in Chapter 5 when we discussed worldviews. However, we *can* take tangible steps to revise these outdated assumptions. Changing our oral and written language is one of the ways of doing this. We need to think about the words we use, and the assumptions and values that could be attached to them. You could follow some of The Writing Center's (2019) advice to keep your writing representative. As part of your editing step, check that you have:

- used gender-neutral nouns (for example, use *chair* or *chairperson* instead of *chairman*; use *humanity* instead of *mankind*);
- avoided modifying nouns related to jobs in order to show the gender of the person (for example, don't write *female doctor, male nurse* or *female scientist*) – indicating gender in these examples strengthens the stereotype that certain jobs should be occupied by a particular gender;
- used surnames and titles in the same way across genders (for example, refer to *Dr Smith and Dr Jones*, and not to *Patricia Smith* [or just *Patricia*] *and Dr Jones*;

- referred to women as *Ms* (not *Miss* or *Mrs*) if you have to include a title that isn't *Dr* or *Professor* etc.;
- used plural pronouns instead of masculine or feminine ones when someone's gender is unknown (*we*, *they*, and *it* are more neutral than *he* or *she* in this situation);
- replaced singular pronouns or subjects with a descriptor or the person's surname (for example, try *Ngubani, the author*, or *the researcher* instead of *he* or *she*);
- used positive descriptions such as *Asian* or *Indian* if you must refer to someone's race or ethnicity (instead of negative ones like *Non-White*, which incorrectly assume whiteness as the 'norm');
- used *sexual orientation* as opposed to sexual *preference* where relevant;
- avoided the use of *he* as a generic pronoun and rather used more than one pronoun if you have to use a singular pronoun and the person's gender is unknown (try *s/he* or *she or he*, instead of *he*).

It has become popular for academic institutions to adopt the plural pronoun *they* as a singular pronoun to replace *she or he* (which can feel clumsy). There are, however, still those who reject this usage because it appears to break traditional English concord rules (i.e. s/he IS, they ARE), so check with your lecturer to see if they are open to you using *they* as a solution (see what I did there?).

In editing for representation, you should also check that you have tried to balance your writing examples across cultures, genders, social classes, world views, abilities, and sexual orientations. For example, you might use a Zulu woman in one example, a gender-neutral English name like Alex in your next one, refer to a female couple in the next, and then to a Venda professor who uses a wheelchair in the example after that (and so forth). The principle is to avoid leaving anyone in your audience feeling excluded, so think about those who might differ from you as part of your writing process – the golden rule here is to make your writing reflective of an equally represented society, as far as possible.

As a final holistic check of this element, The Writing Center (2019) suggests asking yourself whether each person in a diverse group of people who read your paper would feel respected. If you feel confident that this is the case, you should be ready to proceed to the final editing step. You did keep things simple, right?

For a more complete list of checks to perform with regards to keeping your writing democratic and representative, check out The Writing Center's (2019) Gender-Inclusive Language page at <https://writingcenter.unc.edu/tips-and-tools/gender-inclusive-language/>, and the United Nations' Gender-Inclusive Language guidelines at <https://www.un.org/en/gender-inclusive-language/guidelines.shtml>

KISS

As with many areas of writing, that old adage *Keep it Simple, Stupid* (KISS) works very well when editing sentences too. While complex and compound sentences may impress if done really well, it is much safer to reduce unnecessarily complex structures. This means that you need to go through your sentences quite ruthlessly, and cut out 'nice-but-not-necessary' words and phrases, or even entire sentences.

When you do this, look for **redundancy**, which typically refers to information that is expressed more than once, in an unnecessary way.

Some repetition can have a deliberate rhetorical use (see Chapter 9 for specific examples), but repeating ideas or words of similar meaning is simply evidence of poor editing. Tautology is one of the most common forms of redundancy, and refers to saying the same thing but in different words, usually close together. For example, if you wrote:

'In my personal opinion, Theory X is ...'.

you will have included a tautological statement in your sentence. Who else's opinion *could* you be expressing if you said that you were discussing your (in this form 'my') opinion? By using the pronoun *my*, you have already stated the idea – and certainly don't need to add in *personal* too, so your sentence is better as:

'In my opinion, Theory X is ...'

To conclude your editing step, make sure that you have carefully examined your sentence lengths, evaluated and corrected preceding and succeeding sentences, and that your sentences all reflect the simplest, most powerful way of conveying your ideas. If you are satisfied, then you can move on to the very final stage of polishing your writing.

Please visit the Learning Zone to engage with practical editing activities LZ 11.1 and 11.2. Your lecturer may have more activities to offer. By practising, you should dramatically improve your ability to implement these techniques on your real assignments.

11.5 And finally ... proofread

So, you have finished revising your argument, and are convinced that your lecturer is going to be really impressed by it. You have also edited your sentences and think that they lead naturally and logically from one point to the next. You have also removed redundancies from your writing, and are ready to take on that spellchecker. Great!

Unfortunately, proofreading is not quite as simple as that. While it is absolutely essential that you perform a software spellcheck on your work, consider these pointers first, along with a few steps to complete afterwards.

Firstly, check that your spellchecker is set to a suitable version of English. In most southern African universities, United Kingdom English is the preferred option (not US and not South African). Unfortunately, since Microsoft Word is a North American product, its default is US English, so you will need to change this on your computer.

If you need to reset your editing language, perform the following sequence after selecting your entire document (Ctrl+A):

Click the Review tab > Click Language > Click 'Set Proofing Language' > Scroll down until you see 'English (United Kingdom)' and click it > Make sure both boxes below are unticked > Click 'Set As Default'.

Figure 11.2. Proofreading process

Spellcheck

❭ Set up your spellcheck to your specific requirements – UK English is advised.
❭ Accept or reject changes as necessary.
❭ Save your document!

Read aloud

❭ Read each sentence aloud as you visually examine the grammar, punctuation, and language structure.
❭ This helps to spot mistakes that you might have missed if you were just reading the work silently.
❭ Check for typos and for punctuation errors like comma riot, missing full stops, incorrect or incomplete quotation marks or brackets.
❭ Update and save your document each time you do this.

Take a break

❭ Like the break you took before you began your revision, take a break that is long enough to give your brain a fresh perspective, once you have completed your initial reading and corrected punctuation, further spelling errors or typos.
❭ When your brain is refreshed, go back to the 'read aloud' step and repeat it.

'Phone a friend'

❭ When you are absolutely certain that your work is free of spelling errors, typos, and punctuation issues, ask a fellow student – preferably someone who is studying with you, to check your work for any problems.
❭ This is a useful step, but remember that your fellow student is not necessarily good at English, grammar or argumentation, so you still need to take responsibility for the final product.
❭ Address those concerns you think are valid, and save your document again.

Final check

❭ As part of your final check, read through your work once more, this time with a lighter, more removed and objective view.
❭ Check for and correct layout issues, such as headings at the bottom of pages without any content following, incorrect dates.
❭ If you can find no more errors or potential problems, you are ready to submit.
❭ Click save, then print (or better yet, submit online if your university allows this).
❭ For printed documents, check that your layout is still correct before submitting – if you need to correct late layout issues, you will need to reprint.

This *should* set your current and future documents to use English (UK) as your proofing language, so that Word will stop 'shouting' at you in red when you use '-ise' instead of '-ize' at the end of certain words! Just note: in Word, English (US) is sometimes as persistent as a zombie who refuses to die, so make sure you check which language is being used for each assignment. English (US) may have come back from the dead (if so, repeat the above steps).

Once you have sorted out the language that spellcheck should use, check your entire document, and either agree with its suggestions by clicking 'change', or reject them if you wish to leave your writing as is.

Remember that spellcheck is a machine, and it cannot replace your personal attention to detail as a proofreader. All it can try to detect are clear spelling errors, typos, and some of the most regular grammatical and punctuation rules. It does not like complex sentences, and may squeal (with a blue, squiggly line), about split infinitives, even if you *want* your sentence that way to achieve a particular aim. It will also completely miss mistakes due to **homophones** (words that sound the same, but mean different things – such as *their* and *there*), and it may highlight a number of correct words as incorrect, because its built-in dictionary might not be specialised enough.

More importantly, spellcheck cannot understand what you meant to say in your essay, so it will not know if you have used the wrong word as long as it is spelt correctly. For example, if you type 'sue' instead of 'use', or 'your' instead of 'you're', it will not recognise them as problems, so make sure that you check your *own* grammar and punctuation as you run the check, and afterwards.

To complete this process as thoroughly as possible, make sure that you have followed all of the steps in Figure 11.2 above.

 Proofreading is a learned skill – the best way to improve at it is to practise! Visit the Learning Zone for student activity LZ 11.2 for this section. Don't forget to use your skills on your work too!

11.6 Conclusion

This chapter delved into the final stages of reviewing your essay in order to submit a well-constructed and polished product. Specifically, it guided you through the processes of revising your argument structure, editing your sentences, and, finally, proofreading your essay to the point of submission. We hope that by following these processes, the quality of your submissions will reflect the effort made, and that your marks will accordingly reflect this too. Remember, though, that the entire assignment-writing process is important. If these polishing steps appear to be daunting and involved, practising the full process with each assignment will eventually make it a natural part of how you engage with academic essays. It may even take up residence as a mental model that you can apply to your future career.

References

Capote, T. 1985 in Grobel, L. *Conversations with Capote*. New York City, NY: Dutton

The Writing Center. 2019. 'Gender-inclusive language'. *University of North Carolina at Chapel Hill*. Retrieved 31/08/2019, available at <https://writingcenter.unc.edu/tips-and-tools/gender-inclusive-language/>

United Nations. 2019. 'Guidelines for gender-inclusive language in English'. *United Nations*. Retrieved 30/08/2019, available at <https://www.un.org/en/gender-inclusive-language/guidelines.shtml>

CHAPTER 12

Oral presentations

'99% of the population is afraid of public speaking, and of the remaining 1%, 99% of them have nothing original and interesting to say.'
– Jarod Kintz (2011)

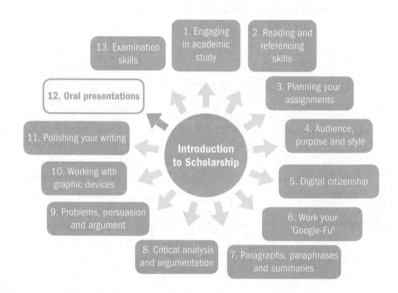

OBJECTIVES

At the end of this chapter, you should be able to:

- use argumentation skills to develop arguments for oral presentations;
- use software such as Microsoft PowerPoint to prepare effective digital presentation slideshows;
- deal with nervousness around presentations;
- use techniques to gain and maintain audience attention;
- integrate visual aids successfully into a presentation;
- deliver a short oral presentation.

IRL

Mbali is really enjoying her first year at varsity. So far, she has adapted well to academic demands. In fact, she has even obtained a number of distinctions! But, as the end of the semester approaches, she feels increasingly anxious about the final examination for one of her modules, for which she will have to deliver an oral presentation.

Mbali is not worried about coming up with a good argument for the topic, since she knows that she has developed excellent argumentation skills since the start of the year. However, she is truly terrified at the thought of speaking in public. This is not new. Her teachers at school used to let her stay after class to do her presentations because she once got so nervous that she fainted. Even then, with just one teacher as her audience, she felt sick and terrified. Thinking about it makes her nauseous. How is she going to get through a speech in front of her classmates?

This time she cannot rely on a special private session – her lecturer has made it clear that all students will present in their tutorial groups, at the given date and time, or receive 0 as their final mark. Although her group is quite small – just 25 students – the thought of standing up in front of them and her lecturer and trying to deliver an interesting speech with a slideshow presentation is absolutely horrifying.

In contrast, her friend Nolwazi is looking forward to the oral presentation. She loves being the centre of attention, and cannot understand why Mbali is so scared. She tells her friend to chill, take a deep breath, and she will be fine. Mbali shakes her head at Nolwazi's advice, annoyed at how easy it appears to be for her, and knowing that it really is more complicated than that.

12.1 Introduction

Some people love the thrill of standing up in front of a crowd, but for many people, like Mbali, the idea of delivering a speech is something that nightmares are made of. Nevertheless, the ability to create and deliver oral presentations is a vital skill for both your academic and working career: regardless of your feelings about public speaking, you need to know how to prepare and present effectively, without falling to pieces and losing your argument in the process.

Using a specific example, this chapter will bring together a number of elements that we have already covered relating to building credible, powerful arguments. It also outlines a useful process for preparing oral presentations. It will offer practical tips on dealing with unruly nerves, and provide you with step-by-step guidance on building solid, effective presentation slides using a popular software package, backed up by the Learning Zone. Importantly, this chapter will also cover how to seamlessly integrate your slideshow into your presentation, together with some key presentation tips, so that you can confidently execute your well-prepared presentation.

12.2 Preparing for oral presentations

Let's assume for the moment that you do *not* suffer from public speaking anxiety at all. In fact, like Nolwazi, you enjoy standing up in front of large groups of people, and charming the proverbial socks off them. Lucky you! But it is not that easy. Confidence and charm might help you get *through* your speech but the majority of the work that you need to do, especially in university and workplace contexts, lies *behind* the charisma. The quality of your argument, and the visual aids that you use to highlight it, are significantly more important than smiling confidently at your audience.

Great news for Mbali and the rest of us who feel rather uncomfortable centre stage: it is *preparation* that makes a real difference in ensuring that you have a quality argument, and in easing nerves.

It is sensible to approach oral presentations the same way that you would approach a written argument, since many of the same principles apply. For example, you need to analyse your audience, fully understand your topic, engage in proper research, and develop a reasoned, convincing argument. But instead of writing it all down in an essay or report, *you* become the product, and present it directly to your audience. Do not panic about this part now – we will help you to deal with your nerves later in the chapter.

Table 12.1 below gives 12 steps to help you prepare for your presentation as thoroughly as possible. Please note that many of these steps include work that you have covered in earlier chapters of this book, so we have included references to where each of these steps is covered in more detail.

Table 12.1. Twelve steps to a successful presentation

Step	Description	Refer to...
1. Analyse the question	Determine what you are being asked to do by conducting a topic analysis on the question. Do this comprehensively to ensure that you fully grasp the nature of the task.	Chapter 3
2. Analyse your audience and purpose	Complete the full audience analysis to ensure that you understand the nature of the audience you will present to, and the purpose of your presentation.	Chapter 4
3. Brainstorm your topic	Determine what you already know and feel about the topic by brainstorming. Let the ideas settle; go back later to start whittling away at what you can use.	Chapter 3
4. Prepare a skeleton outline/ mind map	Use a mind map (or another graphic tool) to organise the viable thoughts that you came up with in Step 3 in order to get a better idea of the areas you need to explore.	Chapter 3
5. Research your topic	Using the basic and intermediate skills that you learnt from this chapter, source suitable, credible material to support your stance, or to build your argument. Take excellent notes, and record the sources that you have used.	Chapter 5 Chapter 6

Table 12.1. Continued

Step	Description	Refer to...
6. Develop a sound argument	*This requires a firm grasp of your thesis statement, the type of reasoning your argument will use, and of your specific claims and warrants. It will require both basic and intermediate argumentation skills, so make use of the GASCAP and Toulmin methods to structure your argument.*	Chapter 8 Chapter 9
7. Revise and refine	*Reflect on the quality of your argument and peel away its 'fluff' by performing a thorough revision. Make use of Aristotle's three rhetorical strategies and the Toulmin method. Be ruthless with the 'fluff', ensuring that what remains is a great argument supported by credible evidence, or, at the least, a persuasive argument with strong reasoning.*	Chapter 9 Chapter 11
8. Create slides or source other relevant media	*Use software such as Microsoft PowerPoint to develop a set of slides – each slide should have minimal content and useful images that will help to guide your presentation.*	Chapter 12
9. Polish your presentation slides and argument	*Spellcheck your slides and thoroughly review your argument – check again for holes in logic, possible counterarguments, and claims with only flimsy support.*	Chapter 9 Chapter 11 Chapter 12
10. Create cue cards	*Use postcard-sized cue cards and number them clearly. Each one should relate to only one claim.*	Chapter 12
11. Practise	*Whether you practise your speech in front of your dog, your sister or brother, or the mirror, make sure you do it! Ensure that your timing fits with your slides or other media. Practice is highly beneficial and helps to polish your presentation.*	Chapter 12
12. Present	*Perhaps the most nerve-wracking part, but certainly not the most difficult. By following a set of tips, you should be able to convince your audience that you are confident of your topic, even if you are terrified. We'll teach you how!*	Chapter 12

As you can see from Table 12.1, the first seven steps of presentation preparation are areas that were covered in earlier chapters of this book, so you should be well-practised in these steps already. This means that you should be able to implement these strategies without much assistance from *this* chapter – it does not mean you can ignore them now. Make sure that you cover these steps carefully before you proceed to steps 8–12; these are covered in detail in this chapter. If you can complete the first 11 of these steps, the actual presentation at step 12 should be a lot less stressful!

To show how these ideas work, in this chapter we will develop a particular topic on the Learning Zone (see student exercise LZ 12.1). For this purpose, we will assume that we were given the following topic:

In early 2015, the Federal Bureau of Investigation (FBI) formally acknowledged that almost all its forensic hair examiners had provided incorrect testimony in trials involving hair comparisons between 1972 and 2000 (Hsu, 2015). This had resulted in 'overstated' positive comparisons that favoured the prosecution in 95% of cases that had already been reviewed, and had therefore resulted in a number of wrongful convictions. A total of 342 reviews had been completed by April 2015, but as a minimum, the FBI still needed to review at least another 1200 cases. Of the flawed forensic cases, at least 32 convictions had resulted in the death penalty, with many defendants being given life or other lengthy sentences.

Calls by human rights advocates to immediately exonerate and release all convicts whose trials were affected by this forensic error have been made. These groups argue that waiting for a full review of all possible cases will keep innocent people unfairly incarcerated for an even longer period of time, thus further infringing on their rights. These groups argue that the only just course of action is to immediately exonerate these individuals.

After researching this topic closely and carefully considering the injustice that has already been done, indicate whether or not you agree with calls to exonerate all convicts who were affected by incorrect forensic assumptions. You should ensure that you are able to support your opinion with relevant data and a sound, reasoned argument.

Using the complete 12-step method in Table 12.1, a response to this argument has been developed, and is available on the Learning Zone as an example (see student exercise LZ 12.1). However, we encourage you to apply these 12 steps to any topics that you need to present on too.

12.3 Developing effective PowerPoint slides

Up to now, you might have endured many lectures, classes or tutorials that involved someone reading in a monotonous voice from a stream of boring, crowded slides. We feel your pain. If so, you may wonder why we would suggest that you incorporate these dreadfully boring, ineffective things into your own presentations.

The reality is that many people who use slideshows use them as a crutch, rather than using them effectively. As a result, their slides become a boring repeat of their argument instead of a useful, visually appealing aid that enhances and guides their presentations. For example, think about how often you have 'switched off' in a lecture because you knew that everything would be on the slideshow handouts? Or, if you knew your lecturer would not give you handouts of the slides, how often have you tried to copy down everything on each over-crowded slide in a slideshow, struggling to keep up with the lecturer as they moved on?

Most of us have been there – whether we are part of the audience or the presenter! After all, nobody *really* teaches us the best way to use slideshows. They are missing from formal curricula. Somehow, we are just expected to be able to use them, and so we go about adapting what we have seen others do, and thank Bill Gates for making PowerPoint so easy to use. Unfortunately, this approach usually does not do justice to our presentations, or Bill's software, and we need to make more of an effort – ironically perhaps, by doing less with the software.

Since the *real* value in using presentation slides is to enhance clarity and understanding, the job of a slideshow is essentially, to:

- amplify your argument;
- speed up the audience's understanding of your argument by providing visual triggers.

12.3.1 Mistakes we make with slideshows

Importantly, slideshows are not there to represent your entire argument on a 2-D surface alongside you. Nor are they there to take your place as the focus of your speech. In this regard, Russell (2019) points out that we tend to make a number of mistakes when we use PowerPoint slides. These mistakes include:

- **Reading slides**
  - > No explanation needed – we have all seen people do this.
  - > If an audience realises the speaker is reading the slides, they may well feel free to go back to playing Candy Crush or whatever else they were doing!
- **TMI** (too much information – overcrowding the slides)
  - > Endless bulleted lists, with sub-bullets and big chunks of text pressure the audience to get through all of the content.
  - > While the audience is reading all this text, they are not listening well.
- **Poor template design** or **theme choice**
  - > Very fancy or arty templates seldom work well, because they are hard to read.
- **Poor colour choice**
  - > Most slideshows start out as white, but this can lead to eye-strain for the audience.
  - > Using many different colours can also confuse the audience, while poor choices of colours can make the text difficult to see.
- **Poor font choice**
  - > Again, fancy or pretty fonts might look pretty when you are sitting in front of a letter, but they make it really difficult for an audience to follow a presentation.
- **Unnecessary graphs and images**
  - > One or two graphs and images might really help to illuminate a speech, but when every slide has a graph, table or image, the impact of the most important ones is lost.

- **Animation station**
  - \> PowerPoint has some amazing animations that allow you to add sounds (like clapping), or use special effects when transitioning from slide to slide, or from element to element within your slides. STEP. AWAY. FROM. THE. ANIMATION. No, really. Some simple animations can help your presentation to move more efficiently, but some presenters go berserk with animations. The audience ends up facing spinning bullet points, mooing cows, and flashing lights. Don't. Just don't.
- **Hardware amateur hour**
  - \> There is nothing that says 'I have no idea what I'm doing' like a presenter struggling with a projector, laptop or microphone after his or her speech is already supposed to have begun. This might be unfair, but if the speaker had time to set things up before and did not take it, the audience may well dismiss what the speaker has to say. Even if this is the *only* weak element in their presentation, it will hurt the speaker's credibility.

Okay, so now you know what *not* to do. Let's establish what you *should* be doing...

12.3.2 Advice for designing PowerPoint slideshows

Watch David Phillips' really interesting TEDx Talk 'How to avoid death by PowerPoint™' at https://www .youtube.com/ watch?v=Iwpi1Lm6dFo

White space is a design principle that refers to everything on the page that is *not* text or image. It helps to balance out the volume of information and avoid information overload. While white space is often white in colour, it can actually be any colour, texture, or background.

An authority on presentations, David Phillips (2014), advises that you use the following principles for developing an *effective* PowerPoint presentation:

- Restrict yourself to one meaningful message per slide.
  - \> Keep your messages simple – less really is more.
  - \> Use a lot of white space, which helps the audience's eyes, and helps them to remember your arguments.
  - \> Silly cartoons or movies do not add much value, so avoid them.
  - \> If you have a part of your presentation when you need your audience to focus on you and do not want them to be distracted by a slide, press the letter 'B' during your slideshow, and your screen will go dark. To restart it, simply press 'B' again, and your slideshow will continue where it left off.
- Use **contrast** – the human eye is attracted to items with high contrast so using high contrast is an effective way to get the audience to focus where you want them to (e.g. in PowerPoint, you can format the points on your slide so that only one shows in high contrast at a time, with the others greyed out or muted).
- **Size matters** – the bigger the item, the more it attracts our attention, so make sure that the key point of your slide is larger than anything else (this includes headings, which are often unimportant).

- If you are speaking while a slide is showing, avoid using full sentences on that slide – either use **short phrases** or stop speaking while the audience digests a full sentence.
- **Avoid a white background with black text** – switch it around. A dark background with light text will reduce the glare of the screen, relaxing the eyes of the audience, as well as encouraging them to focus on you rather than the screen.
- **Stick to a maximum of six points** per slide – using seven or more points requires the audience to use 500% more brainpower to process the information, and their heads will start to ache. This is sometimes referred to by others in the industry as the Rule of Six for slide creation.
  > Limit the title to no more than six words.
  > Limit each slide to no more than six bullets (including sub-bullets).
  > Try not to use tables, but if you have to use them, limit them to no more than six rows.
  > Limit charts and graphs to a maximum of six data points.
- **Use animation and sound sparingly** – a consistent, simple transition between slides is fine, but avoid making words spin or randomly disappear. Transitions can help your audience process information and ready themselves for the next slide, but transitions that are too jazzy are distracting, and look unprofessional.

Note!

Visit the Learning Zone for a practical step-by-step demonstration (student exercise LZ 12.1) of how we used Microsoft PowerPoint to generate slides for our argument on wrongfully convicted individuals in the United States. By following the creation of these slides, from first-click to last, you will see clearly how to adapt your own arguments to slideshow format, and also how to ensure that you have followed the general tips on slideshows in this section.

If you think you have mastered PowerPoint and would like to get more adventurous with your slides, you could try another programme such as Prezi, which offers an interesting set of visual options that help you organise, 'zoom reveal', and engage with your audience more interactively. Prezi is more of an interactive mind map than a series of slides. To get an idea of what you can do with this software, take a look at some of the existing presentations in Prezi's Gallery at <https://prezi.com/gallery/>. Select a topic that looks interesting to you, select full-screen, and then click the right arrows to be taken on a tour through the topic. You will see that Prezi is a fresh, interesting approach to presentations. You can work on it with up to 10 people at once, so it has great potential for group-based projects. Other options you could consider include Visme, Slidedog, Keynote (mainly for Apple users), or Google Slides. Whichever you select, remember to maximise the impact of your delivery by following the general principles discussed in this section.

12.4 Dealing with nerves

Okay, so you have the argument under control, and the presentation slides sorted. But you are still a nervous wreck, with no idea how you are going to pull off your presentation. You imagine yourself walking up to the lecture podium with sweaty palms and armpits, your hands and voice trembling, or worse, mute, then the hysterical laughter of your fellow students and lecturer that follows your five minutes of failure. Are you feeling better yet?

Didn't think so. We really can talk ourselves into a terrible state if we keep focusing on how scared we are, or let our imaginations run away with our 'worst-case scenarios'. This kind of thinking can stem from a fear of failure, or the negative consequences of failure, or might be due to being insufficiently prepared. We may also be worrying about personal insecurities, like our height, size, skin, or voice – things that nobody else is likely to care about, but which feel disproportionately important to us. Being prepared can usually prevent failure, so that is always a great place to start dealing with fear.

However, it is difficult to banish nerves altogether, and to avoid this kind of thinking from ruining your chance at an impressive presentation, it might help to start by understanding a few key ideas about stress itself, because it is stress which drives our nerves.

Firstly, stress is the body's way of dealing with environmental challenges, and can be either negative *or* positive. So, stress can be useful – *if* we do not let it overwhelm us. Try to think about the way stress works as being on a continuum, as in Figure 12.1 below.

Figure 12.1. Stress slide

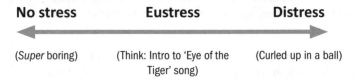

No stress	**Eustress**	**Distress**
(*Super* boring)	(Think: Intro to 'Eye of the Tiger' song)	(Curled up in a ball)

If we imagine stress as a continuum, on the left would be the highly unlikely situation of having **no stress** at all. Even if this situation *did* exist, it would be unimaginably boring, and present no challenges at all. In the middle of the stress continuum is somewhere you would find your 'happy place', called **eustress**, which is a great place to be. This area provides motivation to get things done without terrifying you into a ball of tears and bad decisions. The scariest part of our stress slider is **distress**, at the far right of the continuum, which is a place we'd rather you avoided for your presentation (or anything else). Unfortunately, we all end up in distress on occasion, so it is important to know and recognise it so that you can wiggle your way back to the middle of the continuum.

To put things into perspective for your presentation, it might help to follow the advice of Darlene Price, a communications coach at *Well Said, Inc.* Darlene used to suffer from terrible nervousness – to the point that she fainted during a school presentation! In her attempt to help others overcome their fears, Price (in Smith, 2014) offers some useful insights into public speaking. Perhaps the most intriguing of these insights is that you do not actually need to get rid of your nerves – you simply need to *manage* them so that you can connect and communicate with your audience.

The second important point about nerves is that the audience cannot actually *see* how you feel even if you have a list of terrifying symptoms (unless you hold a big piece of paper in a trembling hand, or faint). Realistically, your audience can only see how you appear and act, so your audience will probably never notice that you are nervous if you learn how to fake appearing calm and confident on the outside! This is really where the idea 'fake it till you make it' comes from – by pretending to be confident, you actually gain confidence, and as your audience starts responding well to your presentation, you gain even more confidence.

So, never fear: successful amateur and professional speakers all agree on several practical ways you can ease presentation nerves. Your (not-so) magical steps to avoiding fainting, blanking and other such horrors, therefore are:

- **Being prepared** (but you're smart and knew this already) – this universal advice means that you need to make sure that you know what you are talking about – feeling uncomfortable with or uncertain about your topic is one of the biggest reasons for feeling distress. Additionally, Best Delegate (2013) says that little acts such as laying your clothes out the night before, printing out your speech cards, and practising your speech can also alleviate anxiety.
- **Checking out the venue** before you present if possible – try to check the setup of the podium, the expected order of speeches, the lights, microphone, or any other tools you might use, if you can. Seeing the venue also helps you to imagine doing your speech beforehand so you can ensure that images and text on your slides suit the size of the venue.
- **Making sure you focus on the message** (your point), **and not the medium** (yourself) – Toastmasters International (2020a) and His (2019), for example, suggest that concentrating on delivering a concise (not waffly), clear message to your audience will mean that you direct anxiety away from your presentation skills (or lack thereof).
- **Imagining yourself in the audience's shoes** – this will help you to accept that most of your audience want to be entertained, engaged, or even informed and challenged. Nobody (okay, sociopaths excepted) wants to watch others fail, so they forgive and forget little mistakes. More often, your audience probably fails to notice them at all. So, little slip-ups might feel like disasters to you, but really are not a big deal to your audience. If you make a small mistake, don't apologise, just move along.

Self-talk refers to the internal 'chatter' that you have with yourself on a daily basis (no, you are not crazy – we all do it). This self-talk can be positive, helpful, critical, negative, or even harmful, so psychologists recommend keeping this inner dialogue healthy and constructive.

- **Knowing a little about your audience's expectations and interests** should help you to find relatable examples for your speech. Additionally, chatting to a few of them informally before the presentation should also help to reduce your nerves and any 'me vs them' feeling you might have.
- **Visualising your success** – okay, this may sound a bit 'new-agey', but experts like Price (in Smith, 2014) and Toastmasters International (2020a) argue that this can really help your nerves, especially if you add **positive self-talk**. By imagining finishing your speech and receiving enthusiastic applause, you immediately relieve some of the negative self-talk that you may have been imagining, and improve your confidence. Similarly, if you tell yourself that you are a brilliant, dynamic speaker, you are likely to act like one. If you convince yourself that you are a terrible speaker, chances are that you will deliver on your worries – so keep positive!
- **Memorising *just* the opening** line of your speech. Ni (2013) strongly advises against memorising your entire presentation – aside from the likelihood of blanking, or delivering it in a boring monotone, it is much more constructive if you understand your topic than if you repeat something word for word. But being confident about the beginning of your speech may ease you into the rest of it and lessen your anxiety.
- **Smiling at your audience** (but not like some Steven King IT clown, please) – Cain (2011) and Price (in Smith, 2014) note that a sincere smile makes you more appealing to your audience, and immediately engages them. Smiling also sends fun chemicals to your brain, which will calm you, helping you to feel less nervous – win-win!
- **Practising deep breathing before you begin** – not a full-on yoga session, but a simple breathing exercise before you begin is a good idea. Ankrom (2020) and Thompson (2017) suggest inhaling slowly through the nose – to a count of four - and then exhaling slowly to a count of five through the mouth. A few repeats should increase the oxygen to your brain, reduce your blood pressure, and alleviate some anxiety.
- **Be yourself** – Nwachukwu (2017), Cain (2011) and His (2019) argue that analysing great speakers and adopting their habits are useful ways to build your presentation skills and appear more confident. But it is equally important that you remain true to your natural style – don't force humour if you feel artificial, and don't try to tell stories if your natural inclination is to be thought-provoking instead. If you are always trying to be something you are not, you are unlikely to feel comfortable presenting.
- **Not equating public-speaking skills with your self-worth** – perhaps the most important bit of advice here is from Ni (2013), who notes that public speaking, while scary for most people, is really just one small thing that people have to do. He argues that even brilliant, capable experts can crash and burn while public speaking, so it is illogical to attach this skill to your value as a human being. Instead, he says it's helpful to see speaking as a skill that can be practised for improvement.

12.5 Delivering your presentation

Your argument and visual aids are ready, your nerves are steadier than they were, and you are ready to stand up and deliver your presentation. With a final few words of advice, your delivery should go off without a hitch.

12.5.1 Use cue cards

You will remember that Step 10 of the presentation process advised you to develop cue cards for your presentation, once you have finished polishing your presentation slides and speech. These are small index-size note cards that contain a phrase or important keywords, which will help to signpost your speech.

We have all seen a speech presented really awkwardly, with the speaker either reading their entire text or holding an A4-size piece of paper in front of his or her face (or both!). Using index cards will help you to avoid both of these traps, and will enable you to make eye contact with your audience, respond to their reactions, make gestures to highlight your point, and move around freely.

The most important benefit of using cue cards is that your speech should come across much more naturally than if you read your entire speech from a piece of paper. Instead of a wooden, rehearsed speech, you should be able to move through your ideas easily, helping you to engage with your audience.

Practical pointers on making cue cards

To create cue cards, you will need a packet of standard index cards (or a few A4 pages cut into 6 or 8 cards) and a thin black pen. Your cards should:

- be clearly numbered on the top or bottom right, and kept in order (you can punch a hole in the top right and tie them together with string if this helps);
- have one main idea on each card (to help you, think back to your skeleton outline or your presentation slides, which feature only the most important ideas);
- use large, clear writing;
- have lots of white space around your text;
- only have writing on one side of them (writing on both sides risks mixing up your cards and confusing the order of your speech order);
- include slideshow prompts to indicate where slides should be changed, or visual aids used – for example, write Slide 2 on the top left of the card;
- on the bottom left of each card, indicate the approximate time you are going to spend discussing each – this way you can track your time. If you are going over time, you can adjust your delivery by leaving out less important points, or add in an extra point if you see that you are getting through your presentation too quickly.

Android and iStore both have cue cards apps that allow you to create, shuffle and edit cue cards using your smartphone as the cue card itself (see *Flashcards App* on Android and *Flashcards+* on iStore). If you prefer to use technology rather than physical cards, this is an option, but you run the risk of your audience feeling as though you are reading from your phone.

Remember that using cue cards needs to be practised. You cannot expect to deliver an immaculate speech immediately after making your cue cards, so spend some time rehearsing first, making sure that you include natural pauses, and that you allocate time for changing slides. You may need to edit your presentation to meet the time limit that you have been given.

12.5.2 Non-verbal behaviour

Non-verbal behaviour refers to the aspects of communication that do not rely on words. It includes gestures, facial expressions, use of time, tone of voice, and speed of speech etc. For example, 'Aidan's jaw was clenched throughout his speech, so it was difficult to hear what he was saying.'

A presentation consists of so much more than just words and an argument. Your non-verbal behaviour provides the audience with a significant amount of information, largely determining whether your audience will pay attention to what you have to say, or not. As a result, it is as important to consider your posture, gestures, speed of delivery and tone of voice, as it is to consider the words that you use to deliver your message.

The first two important non-verbal elements of presentations are that you **arrive ahead of time**, and that you **dress formally**, and suitably for the occasion. Being early will allow you to set up and check equipment before you begin, but will also help you to settle your nerves – you do not want to be worrying about being late on top of worrying about presenting.

In terms of clothing, use your discretion about what you wear, and dress for the audience and purpose of your presentation. If you are an art or design student presenting a bold new idea, then you need not wear the formal clothing that a business management student might expect to wear. For most students, sticking to a dress code of smart shirt or top, and pants or skirt is probably wisest – slightly more formal than what you would usually wear. Remember that your audience's first impression of you is based on the way you present yourself – your clothes and hair are part of this. This does not mean that you need to wear designer clothing, but be smart, and dress neatly if you want your audience to take you seriously.

Some useful tips for positive non-verbal behaviours you might want to try include:

- **Adopting a confident posture** (even if you are faking it) by standing with your back and shoulders straight, and your hands relaxed by your sides. This will open up your lungs, which will help you to breathe more naturally and project your voice (English, 2012). If you are not standing behind a podium, Genard (2019) and Nwachukwu (2017) also advise that you walk as you speak and **'own' the stage** as well. This will allow you to use your whole body to engage your audience – think about how those great TED-Talk-speakers seem to inhabit the entire stage.

- **Making direct eye contact** with your audience (but please avoid a laser-eyed focus on a single audience member for the whole of your speech – that's just creepy). Toastmasters International (2020b) point out that constructive eye-contact establishes a bond with your audience. To accomplish this, Thompson (2017) suggests using an imaginary 'Z' route over your audience members, starting with the back left, and nudging it a little to the right each time so that you make contact with as many audience members as possible for around 4 seconds at a time. This should encourage your audience to pay attention; it also offers you a feedback loop, as you can see if your audience is listening, confused, or engaged – if things are not going well, you can adapt as you go.
- **Using purposeful body language to punctuate your points**. Recall that awesome TED Talk you watched? *THAT*. We do not want you to stand on the stage flapping like a fish out of water, but using your hands, posture, voice pitch and facial expressions can add enormous value to your speech (Genard, 2019; Thompson, 2017). Moving towards your audience during question time might help you to stay connected with them, but avoid annoying, insincere or repetitive gestures and mannerisms (Toastmasters International, 2020b). Experts and average audience members alike are quick to point out how distracting these can be. Fiddling with a pen, pacing, or toying with a laser pointer are likely to distract listeners. Think of ex-president Jacob Zuma's habit of raising his glasses with his middle finger – few people can focus on what he is saying while he does this.
- **Maintaining slow and steady breathing** will help you keep to a natural speaking pace, and allow you to project your voice (Thompson, 2017). As we pointed out earlier, if you did your calming breathing exercises beforehand, and you adopt a confident posture during your speech, you should find it easy enough to maintain a healthy breathing pattern through your speech as well.
- **Using your cue cards as a signpost,** and **not a lifejacket**. Nobody expects you to recall everything from memory, so cue cards are acceptable in most cases. Just don't hold them in front of your face, and make sure that there is only one, short, easy-to-read idea on each card so that a glance at a card is enough. Ni (2013) advises us not to write out our entire speech on our cards then read it out – unless you *want* to send your audience straight to Snoozeville? Didn't think so. When you are not using your signpost-only cue cards, keep them in one hand at your side or on the podium so that they are ready to guide you (but not hide your beautiful face from your audience).

Voice control

The way in which you use your voice to deliver your presentation is an important element of non-verbal behaviour. His hand gestures may be unhelpful, but if you think back to speeches delivered by

Jacob Zuma, you will remember that his voice has a lovely, deep pitch that resonates beautifully. Unfortunately, other aspects of his voice control, like his constant throat clearing, can be distracting.

You should therefore aim for a clear voice, fluent speech, pauses at appropriate places, clear **enunciation**, and variety in pitch and volume to convey your message.

You can begin to achieve these things by standing up straight, holding your head up, and projecting your voice outwards towards your audience. This does not mean shouting at your audience; instead, projecting your words from your diaphragm, rather than your throat, will help to carry your voice effectively without shouting. Professional public speakers and voice experts outline these components of good vocal control that you should consider:

- **Clarity** refers to how clearly you articulate your words – if you remember listening to someone mumble, you know how difficult it can be to hear what they are saying (Mount Holyoke College, 2020). To avoid what Toastmasters International (2011) calls 'mumblitis', try to open your mouth appropriately and give each word time to leave your lips. If you know that you struggle to speak clearly, are shy, or talk fast, doing articulation exercises may help (there are many available online).
- **Fluency** refers to the flow of your speech (University of Minnesota, 2016) and simply means that there are no major interruptions in it. The more fluent you are, the more likely your audience is to be impressed and think you know what you are talking about. Fortunately, the more that you rehearse your presentation, the more fluent it will sound, so practice can definitely make a difference. Remember that being fluent does not mean talking in a continuous stream – pauses are actually essential for conveying meaning. Being fluent means having the right words available when you want them, rather than searching awkwardly for them, or using verbal fillers like *um* and *like* as buffers! Again, practice certainly does lead to improvement.
- **Your rate of speech** refers to how quickly or slowly you naturally speak. While there is no exact rate for which to aim, you should consider your audience, delivery style, and available time. Mount Holyoke College (2020) suggests a range of between 120 words per minute (wpm), which is too slow, and 180 wpm, which is too fast for most audiences. Toastmasters International (2011) suggests that you find your 'happy place' somewhere in the middle, between 140 to 160 wpm, and that you slow down at important points, or speed up to emphasise excitement or humour. If you do this, you should keep your audience engaged.
- **Pronunciation** is an important but sometimes controversial element of speaking. Articulation is the clarity of your words, while pronunciation is speaking those words correctly (University of Minnesota, 2016). Um okay, but correctly according to whom? In South Africa, there is definitely more than one 'correct' English

Enunciation refers to the correct and clear pronunciation of words and their individual syllables.

For articulation exercises, try Write-out-loud's tongue exercises, available at <http://www.write-out-loud.com/tongue-exercises-for-articulation.html>, or watch Vanessa Van Edwards' Vocal Warm Up video that guides you through prepping for speeches and presentations at <https://www.youtube.com/watch?v=7eDcHZZn7hU>. You may feel silly doing these, but the exercises definitely help, so try them out while you are preparing for your presentation, and again on the morning of your presentation.

You might find Oxford's online guide to pronunciation useful for guidance on dictionary pronunciations. It is available at https://www.oxfordlearnersdictionaries.com/us/about/english/pronunciation_english. From this space, you can also type in any word you are struggling with – once you click enter, it will take you to the word, and provide you with a good example of its pronunciation. This is super helpful for those of us who read a lot so know 'big words', but who don't often hear these words spoken out loud.

pronunciation. English is used by many people who speak another language as a home language; many of these people also use English as their academic language. Even those who speak English as a first language do not all share the same pronunciation. The proverbial Queen's English should therefore not be expected in South Africa. Despite these complexities, whichever pronunciation you adopt, it *is* important that you are understood by your audience. For this reason, it is useful to decide on a particular diction – preferably a more formal one for speeches - and that you practise any unfamiliar terms or words before your presentation. If you are unsure of a word's pronunciation (you may have read it, but never heard it said), the University of Minnesota (2016) suggests that you look up the pronunciation in an online dictionary that features pronunciation sound clips. These dictionaries usually allow you to choose a particular pronunciation like British or American English. South African pronunciations (given their variety) are unlikely to be available, so, for consistency, you might want to stick with British English pronunciations.

> **Quick pronunciation tip**: Words consist of syllables, which reflect the spoken, or *sounded* parts of a word. Different languages place emphasis, or stress, on different syllables (Harrington and Cox, 2019). For example, French tends to treat all syllables equally, and so is considered a 'no stress language' (Hyman, 2012), but in English, the emphasis is usually focused on the left side of words, regardless of how these words might be spelt. This is not a hard and fast rule, but as a handy tip, we usually place the stress on the *first part* of a two-syllable word if it is a *noun*, and on the *second part* of the word if it is a *verb* (Ibid.). For example, the word 'convict' as a noun would emphasise the *'con-'*, while as a verb, the stress would be on *'-vict'*. For this reason, if you are stuck on how to pronounce a trickier word in English, and don't have a guide handy, try to identify its part of speech, break it up into its syllables, and then apply this general principle.

• **Pitch**, in this context, refers to how high or low your voice delivers its speech (i.e. is your voice the smooth brandy of Morgan Freeman, or is it more Jar Jar Binks after a fight with a helium balloon?). While this is a personal preference, many listeners prefer deeper pitches, and may underestimate speakers with higher pitches. Although pitch is a physiological feature of your body, we do all have a pitch range in which our voices operate (Mount Holyoke College, 2020), so we can exercise and practise variety in pitch to improve our audience's reception of our message. For example, we can lower our pitch when dealing with important points, or raise it slightly to emphasise questions or uncertainty (Executive Communications Group, 2004).

• **Volume** is obvious: if your audience cannot hear you, then all of your preparation is worthless. It can be difficult – especially if you are shy – to speak loudly enough for your audience to hear you clearly. To help with this, English (2012) recommends projecting your voice from your diaphragm, and not from your throat. Once you have a suitable volume, you can also vary it to attract attention – by lowering your voice at an important point, your audience will have to lean in to hear you, while by raising it slightly, you might better impart a particular point that you feel passionate about. The Executive Communications Group (2004) suggests raising your volume steadily as you build up to a particular point, as it creates a sense of expectation and interest for your audience. If you are using a microphone, make sure that you set it up at the same time as the rest of your equipment: before your speech begins. You could also ask a friend to sit in various seats in the venue while you are testing to ensure that you are audible throughout the room.

- **Vocal variety** puts all these elements together, and is undoubtedly one of the keys to a successful speech. If you agree with William Cowper that 'variety is the spice of life' (2015), then you will probably also agree that a speech delivered in a lively manner, with variety in pitch, tone, pauses, volume, and pace is likely to be far more interesting than one delivered in the never-ending monotone of your most boring teacher ever (we've all been there). Nobody is a 'natural' at vocal variety – it takes practice and comfort with your topic to improve. Try recording yourself to see how you are doing and adjust where necessary.

Integrating visual aids into your presentation

You have spent a lot of time creating an amazing slideshow and developing a strong argument, so it is really important that you use your slides to complement, rather than complicate your argument. To do this, ensure that you:

- **Avoid reading from the big screen** – reading from the big screen will make you turn your head away from the audience, making it hard for them to hear you and losing their attention.
- Remember that *you* **are the presentation** – your slides are just there to enhance understanding.
- **Don't reach upwards and touch the big screen** – use a laser pointing device instead.
- Reveal information at the right time – **don't bombard the audience** with everything at once.
- **Time** each slide 'reveal' at the appropriate point – this should be indicated on your cue cards, so stick to this, and don't get 'trigger-happy' with the clicks.
- For an important slide that you need the audience to focus on, **give them time** to absorb what it says, or prepare the audience for it first, by telling them what to expect.
- Use **verbal transitions** to integrate each slide into your speech – for example, 'As you can see on this chart, the incidences of …'

Barack Obama's 2009 Nobel Lecture is a relatively long speech and does not include presentation slides, but is an excellent example of some of the main principles we have covered in this textbook (argumentation; dealing with counterarguments; rhetoric; and presentation skills), so read and/or watch it to improve your own skills. It is available on the American Rhetoric website in both video and transcript formats, at http://www.americanrhetoric.com/speeches/barackobama/barackobamanobelprizespeech.htm

- If your charts or visual aids need explanation or won't necessarily be immediately clear, make sure that you **explain** them clearly so that the audience understands them in the context of your presentation.
- Avoid providing **handouts** before your presentation – they are like a **'Linus' blanket**, leading the audience to feel like they can ignore your presentation and just read the handout later. If you are required to provide handouts, do so but provide more complete, well-typed and polished notes at the end of your presentation. Remember that your slides are simply highlights, and will not make much sense to your audience when they are at home – notes are much more useful.

In the Peanuts cartoons featuring Snoopy, Linus is almost always seen carrying a blanket while sucking his thumb, so the **Linus blanket** now represents any object that we rely on heavily to comfort ourselves.

Note!

For a comprehensive checklist on preparing and rehearsing your presentation, visit the Learning Zone, which offers an evaluation checklist created using Rate Speeches' Toastmasters Speech Evaluation Checklist Generator. If you can rate yourself favourably (but honestly) using this tool, you should have no problem getting through your actual presentation.

12.6 Conclusion

The tips and tricks presented in this chapter may sound like a lot of work, but it is certainly worth your while to try to prepare your presentations with these in mind. They will help to ensure that you deliver what is expected of you, as well as alleviating some of the stress that you would usually feel before a presentation, building your confidence in delivering presentations in general. As we mentioned earlier, the ability to deliver dynamic, well-thought-out presentations is invaluable, for your studies and in your later career. In this chapter, we covered dealing with presentation nerves, building powerful presentation slides, and developing and using cue cards. Our final section looked at the non-verbal elements of your presentation, and provided practical advice on getting, and keeping, your audience's attention.

References

Ankrom, S. 2020. 'Deep breathing exercises to reduce anxiety'. *VeryWellMind*, updated 3 April 2020. Retrieved 20/05/2020, available at <https://www.verywell mind.com/abdominal-breathing-2584115>

Best Delegate. 2013. 'Attention newcomers – we are here to help: Starting your Model UN career and why nerves are okay'. Retrieved 20/05/2020, available at <https://bestdelegate.com/attention-newcomers-we-are-here-to-help-starting-your-model-un-career-and-why-nerves-are-okay/>

Cain, S. 2011. '10 Public speaking tips for introverts'. *Psychology today*, 25 July 2011. Retrieved 20/05/2020, available at <https://www.psychologytoday.com/us/blog/quiet-the-power-introverts/201107/10-public-speaking-tips-introverts>

Cowper, W. 2015. *The task and other poems*. H. Morley (Ed.). Gutenberg eBook. Retrieved 20/05/2020, available at <https://www.gutenberg.org/files/3698/3698-h/3698-h.htm>

English, J. (Ed.). 2012. *Professional communication: Deliver effective written, spoken and visual messages*. 3rd ed. Cape Town: Juta

Executive Communications Group. 2004. 'There's a message in your voice'. *The total communicator*, 22(3) [online]. Retrieved 20/05/2020, available at <http://totalcommunicator.com/vol2_3/voicemessage.html>

Genard, G. 2019. 'Speak for success: The 5 key body language techniques of public speaking'. *The Genard Method*, 20 January 2019. Retrieved 20/05/2020, available at <https://www.genardmethod.com/blog/bid/144247/the-5-key-body-language-techniques-of-public-speaking>

Harrington, J. and Cox, F. 2019. 'Phonetics and phonology: Syllable and foot: The foot and word stress'. *Macquarie University*. Retrieved 29/08/2019, available at <https://www.mq.edu.au/about/about-the-university/faculties-and-departments/faculty-of-human-sciences/departments-and-centres/department-of-linguistics/our-research/phonetics-and-phonology/speech/phonetics-and-phonology/syllable-and-foot>

His, M. 2019. 'How I conquered my fear of public speaking and learned to give effective presentations'. *Science*, 22 August 2019. Retrieved 20/05/2020, available at <https://www.sciencemag.org/careers/2019/08/how-i-conquered-my-fear-public-speaking-and-learned-give-effective-presentations>

Hsu, S. S. 2015. 'FBI admits flaws in hair analysis over decades'. *The Washington Post*, Public Safety Section, 18 April 2015. Retrieved 30/08/2019, available at <http://www.washingtonpost.com/local/crime/fbi-overstated-forensic-hair-matches-in-nearly-all-criminal-trials-for-decades/2015/04/18/39c8d8c6-e515-11e4-b510-962fcfabc310_story.html>

Hyman, L. M. 2012. 'Do all languages have word accent?'. *UC Berkeley Phonology Lab Annual Report*, 8(1): 32-54.

Kintz, J. 2011. *$3.33 (The title is the price)*. Amazon Kindle. Seattle, WA: Self-published.

Mount Holyoke College. 2020. 'Effective vocal delivery'. *The speaking, arguing, and writing program*. Retrieved 20/05/2020, available at <https://www.mtholyoke.edu/sites/default/files/saw/docs/Effective%20Vocal%20Delivery.pdf>

Ni, P. 2013. '5 Tips for reducing public speaking nervousness'. *Psychology today*, 14 May 2013. Retrieved 20/05/2020, available at <https://www.psychologytoday.com/us/blog/communication-success/201305/5-tips-reducing-public-speaking-nervousness>

Nwachukwu, I. 2017. '7 Tips to improve your public speaking skills'. *ConnectNigeria*, 3 April 2017. Retrieved 20/05/2020, available at <https://connectnigeria.com/articles/2017/04/7-tips-improve-public-speaking-skills/>

Obama, B. H. 2009. 'Nobel Prize for Peace acceptance speech and lecture'. *American Rhetoric*, 10 December 2009. Retrieved 30/08/2019, available at <https://americanrhetoric.com/speeches/barackobama/barackobamanobelprizespeech.htm>

Phillips, D. J. P. 2014. 'How to avoid death by PowerPoint™', TEDxStockholmSalon. *YouTube*. Retrieved 30/08/2019, available at <https://www.youtube.com/watch?v=Iwpi1Lm6dFo>

Russell, W. 2019. 'The 10 most common presentation mistakes'. *Lifewire*. Retrieved 30/08/2019, available at <https://www.lifewire.com/most-common-presentation-mistakes-2767429>

Smith, J. 2014. '11 Tips for calming your nerves before a big presentation'. *Business insider*, 23 June 2014. Retrieved 30/08/2019, available at <http://www.businessinsider.com/tips-for-calming-nerves-before-a-speech-2014-6>

Thompson, S. 2017. '8 Elements of confident body language'. *Virtual Speech*, 10 August 2017. Retrieved 20/05/2020, available at <https://virtualspeech.com/blog/8-elements-of-confident-body-language>

Toastmasters International. 2011. 'Your speaking voice' [PDF]. *Toastmasters International*. Retrieved 20/05/2020, available at <https://www.toastmasters.org/~/media/B7D5C3F93FC3439589BCBF5DBF521132.ashx>

Toastmasters International. 2020a. 'Controlling your fear'. *Toastmasters International*. Retrieved 20/05/2020, available at <http://www.toastmasters.org/resources/controlling-your-fear>

Toastmasters International. 2020b. 'Gestures and body language'. *Toastmasters International*. Retrieved 20/05/2020, available at <http://www.toastmasters.org/resources/public-speaking-tips/gestures-and-body-language>

University of Minnesota. 2016. *Communication in the real world: An introduction to communication studies*. Minneapolis: University of Minnesota Libraries Publishing

Van Edwards, V. 2019. 'Five vocal warm up exercises before meetings, speeches, and presentations'. *YouTube*. Retrieved 31/08/2019, available at <https://www.youtube.com/watch?v=7eDcHZZn7hU>

Write-out-loud. 2020. 'Tongue exercises for articulation'. *Write-out-loud*. Retrieved 30/08/2019, available at <https://www.write-out-loud.com/tongue-exercises-for-articulation.html>

CHAPTER 13

Examination skills

'He who learns but does not think is lost; he who thinks but does not learn is in danger' – Confucius (1963:24)

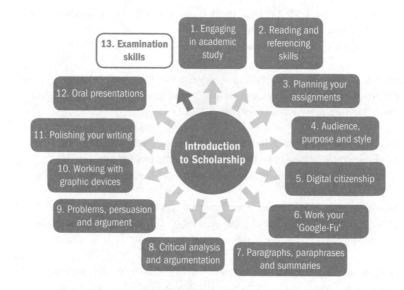

OBJECTIVES

At the end of this chapter, you should be able to:

- use suitable strategies to manage examination stress and anxiety;
- prepare a realistic study schedule for your examination timetable;
- effectively manage the exam process;
- use a variety of study techniques to deal with higher- and lower-order cognitive learning;
- deal effectively with different types of examination question formats.

IRL

Jared and two of his classmates, Hlabangani and Adam, are sitting on the lawn outside their lecture hall talking about their upcoming exams. Jared boasts that the exams are 'Future-Jared's problem' and that since they are two weeks away, they should all just relax. In fact, he asks Hlabs and Adam if they want to go to a music festival for the weekend. Um, no. Thanks. Crazy man.

Hlabangani cannot believe how relaxed Jared is, especially since his marks are right on the borderline, and his friend might fail the semester. Hlabangani's own marks are okay; in fact, so far, they are all up there in the 70s, but he has been having a recurring nightmare lately in which he keeps having to tell his mother that he failed. That dream is *really* starting to freak him out.

He is starting to worry that his nerves will get the better of him in the exam – the first one is just two weeks away. He worries that he will hit a series of 'blanks' – just like he did in his high school biology prelim exam.

Adam, whose uber-methodical approach to studies seems to keep him calm and in charge of his marks, tells Jared that he is an idiot for jeopardising his studies and tells Hlabangani that he just needs a plan to get him through the next few weeks. He advises Hlabangani to draw up a study schedule and to stick to it. At this point, Hlabangani knows he needs to ease his nerves (and his nightmares!) so agrees to try it, although he is somewhat doubtful that a simple study schedule is going to be enough.

13.1 Introduction

Congratulations on making it to this last, but critical, chapter of this book. Your progress probably means that you have reached that part of the semester where exams rear their ugly heads. You need a plan – a workable one – to deal with exam anxiety, the piles of material you need to get through, and the inevitable procrastination that comes with assessments.

Before we start dealing with all of that, it might be helpful to remind yourself *why* you need to write exams. You might suspect that exams are there to ruin your social life for month-long stretches at a time, but you have probably already figured out that they are part of the way your lecturer determines whether you pass, fail or excel at a subject. Exams also serve some other valuable purposes.

Firstly, your institution's reputation depends on the quality of the graduates it sends into the working world. If an institute sent out unprepared graduates, its reputation, and of course its students' ability to find employment, would be jeopardised. So, your university uses exams to check that you are ready for the career area you studied for, keep its reputation up to scratch, *and* make sure that the job market wants to hire its graduates. Exams therefore are a win–win situation for both university and graduate.

Secondly, by writing exams, you provide your university with a great opportunity to evaluate the quality of its courses – exams allow your institution to note strengths and weaknesses in its courses and in student comprehension, which it can then act on for future courses. For example, academic literacy courses, covering the kind of material in this textbook, are the result of lecturers noticing skills-gaps in students' exam answers that arise out of the move from secondary to tertiary. If students did not write exams, it would be difficult to find out where they were struggling, so these are a major benefit to course design.

Thirdly, writing exams provides you with a vital opportunity to reflect on what you have learnt, and allows you to piece together everything that you have done in order to generate a more cohesive picture of the semester's work. In other words, exams actually encourage you to think critically about what you have covered, and on how much, or little, you have understood about the course. This is actually one of the most valuable tools that you might gain at university. We encourage you to keep this in mind as you study for your finals – you should benefit much more from your course as a whole if you avoid parrot-fashion studying, and strive to understand and contextualise your work.

Okay, but what if you are like Hlabangani in our IRL scenario – worrying that you will have an exam 'blank'? In this case, it would be useful to consider the observation of Roger Mead, a stress management consultant, (in Duff, 2019) that when it comes to exam stress, it is the way that we *perceive* exams that is bad, not the exams themselves. In fact, our perception of exams is often so bad that we develop major exam anxiety.

An important study (Cassady and Johnson, 2002) on examination anxiety claims that high levels of this anxiety can play a negative role in the study process, and on the assessment itself. This is because anxiety can cause you to organise or conceptualise information poorly during study time, and so limit your ability to recall important information, or to present it coherently during the actual exam (Ibid.). It is this element that clearly bothers Hlabangani, and he needs to work through it.

If you can approach your exams in an organised way in a reasonable amount of time, you should reduce your anxiety considerably. This chapter will help you to synthesise some of the important skills from previous chapters into a practical approach for dealing with examinations. In addition to the goals and attitudes that you reflected on in the first chapter of this book, core examination skills include your ability to analyse questions, use different types of reading, write paragraphs, summaries, paraphrases and arguments. You will need to ensure that you are well versed in all of these.

If you have been following the chapters in this book and completing the activities on the Learning Zone, you should be able to comfortably apply your knowledge to the exams, which are simply an assessment of the content and skills that you have covered already. If you haven't, um ... start reading and doing, please!

Okay, so you have got the core skills down, and are ready to synthesise them for your exams. In addition to these skills, we need to address anxiety and procrastination, two of the most common issues around exams. This chapter will therefore also help you to deal with exam stress by providing simple but effective tips, as well as offering practical advice on avoiding procrastination and using your time effectively. By helping you to develop a realistic exam schedule and guiding you through various approaches to different types of exam questions, we hope that your feelings about any upcoming exams will be a lot more positive, and that you will find the best methods to help you do well!

13.2 Dealing with stress and anxiety

Unless you are a unicorn, you will have experienced anxiety, whether it was standing up to speak in front of your class, speaking to that person you fancied for the first time, or even during your finals at school. Your hands were clammy, your mouth was dry, your heart was rabbiting uncontrollably, and your stomach felt like someone was wringing it out like a wet towel. We get it: life is full of anxious moments. Some of us have figured out ways to deal with anxiety most of the time, but still experience it at other times; the rest of us are nervous wrecks when faced with almost any new or important thing we have to deal with. The reality is that we all have to know how to cope with anxiety if we want to succeed at new or difficult tasks, including exams.

In Chapter 12, we briefly introduced the concept of stress, noting that if we think of stress as a continuum, we can find a 'happy place' of **eustress** somewhere in the middle. Eustress offers us the right amount of motivation to move forward: not so much stress that we are debilitated by anxiety, but enough stress to avoid the kind of boredom that includes watching infomercials. In other words, we want to find a place of eustress, and avoid feeling distress or an absence of stress.

In our IRL scenario of three friends above, only Adam seems to have found this 'happy place' for studying. His methodical approach is working for him, his **academic identity** looks well-developed, and he seems to be managing his stress levels well. Ultimately, this is where we would like to see you working too!

Hlabangani, despite his good marks, is starting to feel distress, which is clear from his nightmares and his current thinking. If he does not move away from this, he could perform poorly, even if he tries hard to prepare well. This would probably be because his nerves got the better of him, as the Cassady and Johnson (2002) study suggests, not because he is unfamiliar with the work. He still has two weeks before his first exam and his marks are good, so Hlabangani has time to move towards eustress if he adopts some of the advice that Adam gives him and manages to calm down.

Jared, who *appears* to have no current stress, is more than likely just 'ostriching' by thinking he will get through everything in 'Future-Jared's' time-zone. Although his marks are only so-so, he seems to think that studying just before his exams will get him through. The reality is that when he opens his books the day before his exams, he is likely to be horrified at the work he has to get through. He may try to race through his work, spot-study, or simply give up. Regardless of what he does, he is likely to move into distress and high anxiety. He will not have time to correct this, so his chances of succeeding in his upcoming exams are, quite frankly, dismal.

Before you form a plan of attack for your exams, reflect on your current levels of anxiety. Visit a stress test, available at https://www.psychologistworld.com/stress/stress-test. After completing it and reading the assessment of your stress levels and their suggestions, decide whether or not you agree with the test's findings. If you do not agree, do you feel that you are more or less stressed than the test said you were? Think about why you do not agree. If possible, discuss this with a classmate who has also taken the test.

If you have followed the advice offered in Chapter 1 of this book, and have diligently taken notes, completed your revision after each lecture, and prepared fully for your classes, you should pat yourself on the back for saving yourself a significant amount of worry – if you have engaged with the work throughout the semester, most of your exam preparation is already done. All that is left now is to revise what you have covered, organise it into logical components, and then ensure that you are familiar with what you need to be able to do in the exams. Even so, you might still feel a little nervous.

Hall-Flavin (2013), a medical doctor who consulted for the Mayo Clinic, says that some nervousness before assessments is perfectly normal. But anxiety that makes you doubt yourself and therefore potentially impacts on your performance should be managed. He offers a number of helpful strategies – most of which are common-sense – that you can use to try to reduce your levels of anxiety. Table 13.1 below combines Hall-Flavin's (2013) more general advice with some practical relaxation techniques. You may feel silly doing these at first, but try to find a quiet, private space so that you can attempt them without feeling judged! Not all of these may work for you, but you should be able to find at least one that helps you to relax.

Hall-Flavin (2013) also suggests a number of other strategies that you can use to reduce your levels of exam anxiety. These include:

- Creating and maintaining a **consistent exam routine**
  - > When you find a routine that works for you, try to follow it for your other exams – this helps to ease your nerves about the unknown.
  - > For example, if you like to wake up early on the day of the exam, have a good breakfast, and then get to the exam venue ahead of time so that you get yourself settled into your seat in plenty of time, keep doing this. If you realise that speaking to your classmates just before or after the exam makes you more anxious, avoid doing this for future exams.

Table 13.1. Relaxation techniques for reducing exam anxiety

Deep breathing technique	• Lie flat or sit comfortably on a chair with your back straight – place one hand on your chest and the other on your stomach. • Breathe in through your nose for 3–4 counts – you should feel your stomach move, but not your chest. • Exhale from your mouth for as many counts as possible, pushing out all the air you can while squeezing your stomach muscles – again, you should feel your stomach move, but not your chest. • There is no limit on how many of these you do, but try at least 20 deep breathing sequences before stopping.
Progressive muscle relaxation	• Lie on the floor in a comfortable position, making sure that you are wearing loose clothing. • Take some deep breaths, as described in the deep breathing technique to relax yourself a little. • Starting with your right foot, alternate slowly tensing each of the muscles in your body as tightly as you can on a long in-breath with relaxing each one on a long outbreath. Perform this from the bottom to the top of your body. Start by tensing then relaxing your right foot, then move to your left foot, right ankle, left ankle, right calf, left calf, etc., all the way up your body to your face. Note: ensure that you are only tensing the muscle that you are working on! • Once you have relaxed all the muscles in your body, stay in your position, and use a deep breathing technique for a while until you feel completely relaxed.
Visualisation meditation	• Find a quiet, relaxing space, and sit in a comfortable position. • Close your eyes and imagine a place that you find very calming. • Picture your calm place as intensely as you can by using your senses: imagine how it would look and smell, what it would feel like, what sounds you would hear, and what tastes you would experience in your place. • Imagine yourself slowly walking around your chosen place, noticing the colours and textures, letting your senses move around. Taste the water or fresh air, inhale the smell of flowers or aromas of cooking, listen to the gentle sounds of water running, or birds chirping, depending on the space you have chosen. • Gently open your eyes when you feel relaxed. • Note: don't worry if you zone out during this exercise – it's quite normal. You should feel quite relaxed afterwards.

• Ensuring that you are **eating and drinking suitable things** in suitable quantities
  > Your brain needs a good deal of healthy food to function at its best.
  > Avoid energy drinks, fizzy cooldrinks, or coffee (sad but true), since caffeine can increase feelings of anxiety.
  > Drink loads of water, and snack on fruits and nuts rather than crisps or fast food.
• **Getting sleep**
  > You might feel the need to study through the night, but your body needs sleep, and your brain needs time to digest all the content that you have been trying to stuff into it – if you deprive your brain of the sleep it needs to process the content, you may not be able to recall the information later when you need it, so your 24-hour session will have been wasted.

- **Getting exercise**
  - > Do not attempt a marathon, but do try to get some fresh air into your lungs by doing regular exercise, especially on the day of your exam.
  - > Exercise also reduces tension levels and frustration, so it can help to clear your mind and allow it to focus on more important things.
- **Asking for help** if you need it
  - > If you are struggling with a special need like dyslexia, or with something more general like not knowing how to study properly, don't ignore it – failing to deal with a problem is likely to make you even more anxious.
  - > Speak to your lecturer or the people on campus who are equipped to help you with these issues, and try to find a practical way around them. For example, you could get extra time to complete assessments, or be shown a better way of studying, depending on what you need.
  - > Remember to keep things in perspective – an exam is just *one* part of a small phase of your life – it should not threaten to overwhelm your happiness. Ask for help if you need it. If you feel significant anxiety, or really overwhelmed, please visit your campus counsellor, who is qualified to assist you, and who should have a good understanding of the problems you may be experiencing. Don't keep quiet!

Point of interest

If you don't feel comfortable talking face-to-face with your campus counsellor, friends, classmates, or family, you can always contact the South African Depression and Anxiety Group's (SADAG) Suicide Crisis Line, which is available at the toll-free number: 0800 567 567 or via SMS at 32312. They can help you deal with depression, anxiety, substance abuse, sleep disorders and other issues.

Ultimately, preparing adequately and having a clear idea of what you need to achieve in the time you have available are key to feeling confident when approaching exams. As a result, learning to study efficiently will help to significantly alleviate your stress levels, by managing your time effectively and having a solid, realistic schedule.

13.3 Time management

There are **three levels of time management** that you need to use to prepare well for exams. This does involve effort, but it does not make sense to risk your study fees, the effort you have already made, and the extra time needed to redo a subject by not actively managing your time. So, get on top of things by following some sound time management advice, and you will not end up looking like Surprised Koala!

We introduced you to **long-term time management** in Chapter 1, and hope that you integrated it into your study manifesto and studies from the get-go. You are already in a great position if you have prepared for each class and continuously revised your work throughout the semester. If you have completed all given tasks, activities and studied for the formative assessments that you have been given, ensuring that you have understood all of the work covered in the semester, you have little to worry about, since these are major components of what you need to address for your exams.

Generated on https://imgflip.com/memegenerator

If you have let your long-term time management slip during the semester, you need to recognise that your job preparing for the exams is going to be a lot more intensive, and is going to require more effort than those who have kept at it. Don't freak out though – if you still have a couple of weeks to prepare, it is not the end of the world. It will be tough, and it might be difficult to get excellent marks, but if you focus your energies on short-term preparation, you *can* still be ready for your exams.

There are a few very important areas of **short-term time management** that you need to think about, whether or not you have kept up with long-term time management. These are:

1. developing a realistic **study schedule** (see Section 13.3.1) a few weeks (but preferably a month) ahead of your examinations;
2. **sticking** to your study schedule and avoiding procrastination;

3. finding **previous exam papers** for all of your modules, and then familiarising yourself with the style of questions so that you are prepared for the general approach of your exams;

4. making sure that you consider any **exam briefs** that your lecturer may provide – if these are available, they usually place clear limits on the volume or nature of work that you need to study, or give you other valuable advice. Of course, you need to be in class to catch these (they may be written or spoken), so keep attending near exam time – this is often the period that your lecturer provides useful information on the exam!

The examination itself is the final bit of time that you need to manage effectively. **Examination time management** means that you need to ensure that you:

- get an overview of the general format, instructions, types of questions, and mark allocations of the examination during the **reading period** – this will help to settle your nerves about the previous 'unknown';

- decide **which questions to approach first**, based on the reading period, in order to maximise your marks – identify those where you are really sure about the answers and start with them. Just remember to label your answers clearly and correctly so that the marker knows which question s/he is dealing with;

- provide a **sufficient response** for the mark allocation, but avoid an excessively long response that will not get you more marks – in general, a mark-per-point or mark-per-minute principle usually applies, unless the exam paper or lecturer says otherwise. Make sure that you note the mark allocation of each question and stick to it!

- respond *only* **to the specific question asked**, and not to what you wish had been asked. This means that you should analyse the question (see Chapter 3) to respond to the direction word (verb) and the specific question area itself. If your answer includes irrelevant information, you will not be awarded marks for it, even if the content is correct. Answer the questions you have been given, at the appropriate cognitive level for the instruction verb. For example, if you have been asked to *debate*, simply *describing* something is not enough – but on the other hand, if you have been asked to *list* something, do not waste time *detailing* it!

- allow time to **polish and revise** your examination responses and to make sure that you have answered all questions as fully and correctly as possible. Watch that you do not introduce errors or second-guess your initial answers in this stage – if you are sure that an earlier answer was incorrect and know the right answer, change it, but do not change initial answers if you are unsure.

There is one last period of time you should also consider: the short period immediately following the exam. Students often like to compare notes after an exam, dissecting it to establish who answered which questions in which way. This is a bad idea. You might have felt confident immediately after writing the paper, but then feel suddenly doubtful after a 'dissection session'. Instead realise that your classmates are likely to have responded differently to questions, and that their responses could as easily be wrong as right. If you find out that your answers are different to those of your classmates, you may feel anxious, to absolutely no benefit at all. Until your actual results are released, you will not know how you did, so avoid these fruitless discussions, focusing instead on preparing for your next exam.

13.3.1 Developing a realistic study schedule

Firstly, remember that no one can prescribe a particular approach for organising your time. Different students learn things differently, and take different approaches to their studies. Your best friend may take a whole day to learn a few sections of a subject, while you might fly through it more quickly – or perhaps more slowly. So, you need to *realistically* assess how you study, and try to determine how long you think it will take *you* to prepare properly for your exams.

If you have managed your time in the long term correctly, and still have a few weeks to go until your exams, it should not be a problem for you to develop a realistic study schedule. You should be able to adapt by including a few extra hours a day for revising and exam preparation. If your social or Google life was too important during the semester, and you did not manage your time in the long term, you will need an intensive period of study and revision.

Figure 13.1 on the next page outlines a step-by-step process that you can use to create a study schedule for yourself. We have used military terms for this process, because if you apply military discipline to the process, you might stick to it better!

When you are allocating time to individual tasks, remember that humans tend to underestimate the amount of time we take to accomplish things (see Chapter 8). For this reason, you should make sure that you are allowing realistic time slots when you are developing your plan – rather overestimate than underestimate. You do not want to realise the day before your exam that you only managed to cover half your work despite sticking to your plan!

The website *Intelligent* is a useful online resource, offering study advice with practical tips, as well as links to templates that you might want to use. You can access their page at <https://www.intelligent.com/create-a-study-plan/>. Just make sure you don't get 'lost' surfing the links and end up wasting more time than you save!

Figure 13.1. Creating and using a study schedule

Reconnoitre

> In the military, to reconnoitre means to gather all relevant information - the same applies here!
> Think about your personal study-time preference (no, 'never' is *not* an option). Do you study better in the afternoon, evening, or morning?
> Identify all responsibilities that you have – these include chores, your job, exercise, grocery shopping, important social events (try to limit these to important ones only), and, most significantly, *all* your individual subjects, any regular homework or assignments still due.
> Source the correct dates and any other relevant details for all of your events and write these next to your list of responsibilities.
> Decide on a format for your schedule – you can use a spreadsheet or table format, or one of the many that are already available online (e.g. Sara Laughed for some excellent study planning templates, available at http://saralaughed.com/free-library/). Hand-drawn tables are fine, but they are difficult to change, are hard to re-use and will take a lot longer than a digital version.

Complete the plan

> Depending on how far in advance you are planning (i.e. daily, weekly, monthly), complete your study schedule by populating it with all of the responsibilities and dates that you sourced in the reconnoitring phase – remember to consider the time you find most suits your 'study-brain' and try to schedule study sessions for that time of day if possible.
> Be realistic – a plan to work for 16 hours a day might look impressive, but is not really doable. Instead, plan shorter work sessions and a realistic daily total.
> Try to block your study times if possible – spending a few hours on one subject might help you to get into the 'swing' of it more easily than working only for half an hour.
> Schedule more time for more difficult subjects – and try to mix these study sessions with sessions on easier subjects so you keep your motivation high.
> Be as specific as possible – writing 'Chapter 1 & 2 PSYCH101' gives you a more visual goal than just 'PSYCH101', which leaves you with wiggle room.
> Make sure that you schedule breaks – each person is different, but 5-15 minutes per 45 minute or hour is generally helpful.
> Use colours to categorise your items, e.g. study time vs relaxing time, or different colours per subject, so that your schedule is easier to follow.

Execute your mission

> Make sure that you keep your schedule handy – it should become part of your everyday routine.
> Stick to your study plan as far as possible but do not panic if you need to shuffle things around though – life happens. Just make sure that if you need to skip a session, you make time for it at a later stage.
> Revise your plan if you see that you have under- or overestimated any areas of your plan. Remember that this is purely for *your* use, so it needs to work for you.

13.3.2 Procrastination station – step away from the Google rabbit hole!

We have all been there: an exam in two days' time, and absolutely no time to spare, and yet, here you are, lost in the fascinating world of exam memes. Again. You tell yourself to move along, but when you look again, another hour has gone by, and you have found your way to *The Website that Time Forgot*. When you think about it, your procrastination is illogical: you are studying something that you find interesting (we hope!), and actually enjoy finding out more about your subject – yet here you are actively trying to avoid it.

Mark McGuinness (2020) says that 'if it weren't for procrastination, we'd all be superheroes', so why do we fight so hard to avoid doing what needs to be done? He argues (Ibid.) that we sometimes procrastinate because we do not know *what* to do, and at other times because we do not know *how* to do it.

However, when it comes to studying, these are seldom the reasons for delay: usually we know *what* we need to do and *how* to do it – we just resist doing it! McGuinness (2020) says that in this case, the reason for our procrastination is **internal resistance**, which arises when we try to move outside of our comfort zones in order to face a challenge. Consider what Pressfield (in McGuinness, 2020) says about resistance: it actually separates the professional from the amateur – where an amateur gives in to resistance (and promptly proceeds to another game of Candy Crush), a professional perseveres through the doubt and tries anyway. The simple argument is that if we allow resistance to win, we will not get to be superheroes.

How *do* we overcome our resistance, our tendency to procrastinate? How *do* we develop a bullet-proof academic identity? In the next section, we provide some practical tips for avoiding procrastination (adapted from Chua, 2019), so that you can get on with the business of becoming an academic superhero!

1. Firstly, you need to **be enthusiastic** about what you are doing – if you are interested in your subject and bring a positive attitude to it, it will be much easier to engage with it, even if it is cognitively challenging! If you are unenthusiastic or bored by your subjects, try to find something that you *do* find interesting in your subject – try to change your perspective. For example, if you do not like English poetry but do like maths, look at the mathematical patterns in the metre of poetry – you might start to enjoy English poetry more.

2. Make sure that your **study space is organised**, and well set up – if it looks more like a gamer's paradise, you are going to end up doing precisely that. If it is way too comfortable with loads of fluffy blankets and a big TV, you will end up watching reruns, curled up with coffee. You need to be set up for studying: have a clear desk space with plenty of coloured highlighters, note-paper, a dictionary, your relevant textbooks and notes, a computer for looking up any troublesome elements of your work (only use it for that!), and a comfy (but not too comfy) chair that allows you to sit up straight. Your room needs plenty of natural light, or enough artificial light to allow you to spend time reading without having dealing with glare. Keep some healthy snacks and water handy so that you do not need to interrupt a study session. Importantly, make sure you have organised your space at least two weeks before your exams – otherwise this step could act as another procrastination tool!

3. **Use the study schedule** you created, making sure it is easily visible. Tick off items that you have done so that you track your progress. Remember that you are less likely to drop the ball if you are specific and break up your studies into manageable chunks. You are less likely to jeopardise your plans if you know that you have many small steps to complete rather than just one monster deadline.

4. Revisit your **personal study manifesto** that you created for Chapter 1. You should have updated it throughout the semester, so now you can use it to remind yourself of the goals you set. Adapt what needs to be changed – for example, you might find that you had aimed just to pass a particular subject, but now feel that you would like to work towards a higher mark in it, or that another subject is actually more difficult than you thought, so to be realistic, you need to lower your expectations. Regardless of what you need to tweak, make sure that you keep your overall goals in mind all the time – this will make it more difficult for you to slack off!

5. **Tell everyone** who will sit still for five minutes what your goals are. There is nothing like a bit of positive peer pressure to keep you working towards your goals. Those goals we set for ourselves that remain private tend to be easily swept under the carpet. But if you tell everyone that you aim to do X, Y and Z, they are likely to ask you about your progress, and therefore keep you working towards them.

6. Find a reliable **study-buddy**, and hold each other accountable for meeting deadlines, and working towards goals. The shared 'pain' of studying provides an excellent support structure, and many long-term friendships form in this way.

7. Associate with **people who inspire** you, and avoid those who seem content with the status quo – if you do not personally know anyone who inspires or motivates you, visit the websites of inspirational figures who seem to get up and do things that they put their minds to (but don't get stuck there, please!). Enthusiasm and positivity is infectious, so individuals like this may help you to focus on your goals and keep moving forward.

8. Try to **contact graduates** who have already passed your subjects – tutors are usually a good choice here. They can advise you well, and will also provide you with living 'evidence' that it is possible to get through the work, even if it currently looks impossibly difficult.

9. **Don't wait – just do it**. One of the biggest reasons we procrastinate is because we are waiting for the perfect time to begin. That 'perfect time'? Not going to happen. We will *always* have other things that we are busy with, so just get going and start studying. Instead of saying 'as soon as my assignments are all done, I will start my exam prep', think about this – how many hours will you have wasted not preparing for your assignments, because you waited until the last minute to do them anyway, *and* then ended up wasting study time too? The key is to stop making excuses, and

instead get on with things. Studying is hard work. We know that. Now get on with it. Once you have started, it will be easier.

10. This might seem strange: avoid making 'study' notes as a form of procrastination. This is a sneaky, self-delusional tactic that we engage in when we decide that we need 'proper' study notes. We spend hours making them look pretty and ensuring that they include everything possible. The only problem is that we are more concerned with the way that they *look* than actually thinking, reflecting, and understanding the content. Taking notes is a valuable form of revision *if* we think about what we are engaging with. Proper study notes will never include all the information that you need to study – it will include broad concepts and ideas, using key words. When we are simply transferring information from one textbook to our notes in a pretty way, we are trying to fool ourselves that we are busy with the work itself, although we are actually procrastinating. Talk about an own-goal!

11. The most important way to avoid procrastination may be to **get rid of temptation!** If you know that you are a social media junkie, put your phone in a different room, or remove any social media 'favourites' tabs if you are using a computer to study. That 'quick' check to see if anything mind-blowing has come up can turn into a very long session, so don't do it. Avoid doing things in your 15-minute breaks that you know might lead to a longer break – telling yourself you are going to watch television, or play an online game for 15 minutes is a self-lie. You know it is never going to be just 15 minutes, so don't do it.

13.4 Study tips

If you make use of critical and study reading – the **SQ3R method** as covered in Chapter 2 – you should glean a solid understanding of the content of most of your modules. Combined with **revising past papers**, this, and similar reading strategies, should allow you to work with the theory that you have covered, and apply it to any questions or scenarios that you are given in the exams. Although this study reading technique is not covered again in this chapter, it is the most valuable approach to your studies overall. Critical study reading provides a much deeper understanding of your coursework than simply trying to memorise content, and allows you to 'connect the dots' between theory and practice. Therefore this method should be the core of your study technique.

If you examine past papers, you will notice that some areas of your work *do* require memorisation. And unless you are lucky enough to have a photographic (or eidetic) memory (which science says does not exist anyway, see Weller, 2014), this will mean using a few tricks to ensure that you *can* recall the little bits and bobs of information.

There are hundreds of mnemonic rhymes and words; you can also create your own! For some well-known examples, visit ThoughtCo.'s page on mnemonic devices for students at https://www.thoughtco.com/mnemonic-devices-tools-7755

Mnemonic tricks are memory devices that help you to recall larger bits of information. Pronounce mnemonic as though it starts with an 'n' not an 'm': nim-**on**-ik.

Note that there is a difference between remembering and recalling – your brain might remember (or store) information, but it is useless to you if your brain cannot recall (locate and express) it during exams. For this reason, if you are sure that you understand the content, and now need to simply train your brain to recall it in your exam, these mnemonic tricks may help:

- Using nonsense words or sayings made up of the initial letters of the items you need to recall, for example:
  > HHeLiBeBCNOFNeNaMgAlSiPSClArKCa – a ridiculous word made up of the abbreviations for the first 20 elements of the periodic table. Ridiculous as it is, it somehow stays in our heads, and has helped many science students to remember the elements;
  > **N**ever **E**at **S**ilk **W**orms (the points of the compass in a clockwise direction from North, to East, to South, to West). Silly, but when we look at a compass, this rhyme still pops into our heads;
  > In mathematics, the **BODMAS** rule, which dictates the order in which you should proceed with a calculation (**B**rackets, **O**rders, **D**ivision or **M**ultiplication, **A**ddition or **S**ubtraction);
  > **K**ing **P**hilip **C**uts **O**pen **F**ive **G**reen **S**nakes, used in biology to remember the order of taxonomy (**K**ingdom; **P**hylum; **C**lass; **O**rder; **F**amily; **G**enus; **S**pecies).
- Using rhymes of your own, or existing ones, like 'Thirty days hath September, April, June and November …' for the number of days in the months;
- Making **numbered lists** does not fall under mnemonic devices, but lists can help if you prefer to learn in a linear fashion – by attaching a number to a point, you help to sequence the points that you need to remember, and this helps you to recall as many items as necessary;
- Using vertical or horizontal **timelines** can also assist you to recall events in sequence – if you think about each of the items on your timeline as a step to the next item, you should recall the information more readily, even if you have to trace your steps back;
- Making **bulleted lists** is probably the most common way of making notes – indenting sub-points and underlining headings can help your brain to organise information logically, and therefore help you to recall it when necessary.

Other strategies include using **Post-it notes** of key concepts placed in strategic locations in your home or study area; creating giant posters of **summary tables** for your wall or ceiling; using large **spider diagrams**; discussing concepts with a **study-buddy**; or engaging with your class tutors or your institution's online Learner Management System. The more you expose yourself to the content, the easier it becomes to recall.

13.5 Active listening skills for class

Have you ever struggled to answer a question in an exam and thought to yourself: 'I recall the *exact* day and time my lecturer discussed this in class, and even remember that stupid yellow tie he was wearing – why can't I remember what he actually *said* about this?' Perhaps you have been in a heated debate and realised that you were so busy planning your next point in your head that you failed to hear the other person's point. Or maybe you were frantically taking notes in class, but when you read them later, you realised you had not actually understood what your lecturer was saying. If you are 'guilty' of any of the above, don't feel alone: we all have our moments (or even habits) of inattention, and have all experienced some of the consequences that come of this in our academic, personal, or work lives.

In Chapter 1, we referred you to this section if you felt your listening skills were not up to scratch – congratulations if you admitted this early on and headed here! Many students do not realise that they *can* (and probably *need* to) improve their active listening skills, because they assume that listening is the same as hearing. But hearing and listening are quite different things: **hearing** is a physical process that refers to the *detection and translation of sound* waves by our ears into signals our brains can interpret; it happens automatically (Martini, 2005). On the other hand, active **listening** requires intentional, motivated focus and is necessary for us to understand *the meaning* of whatever we have just heard (Caspersz and Stasinska, 2015). Listening *uses* hearing, but it requires far more than that. With some guidance, motivation, and patience, you *can* develop your active listening skills – even if they are dismal now – to ensure that you are equipped for academic (and social) listening activities that you encounter.

Right how do we go about doing this? There are a number of different types of listening, but let's start with developing key **active and critical listening skills** in the context of a lecture:

1. Firstly, **attend** any scheduled classes, tutorials, or online lectures is a must (obvs). If you are present, you get the opportunity to hear, interact, and engage in any activities – through these, you should learn a lot more than if you stayed home on your couch playing Fortnite.

2. Secondly, make sure that you **prepare** properly for class – identify your goals for the class beforehand. For example, if you need to read specific chapters or articles for a class, make sure that you do. Check your study guide or course objectives to help you identify what the likely objectives of the particular lecture will be, and focus on these areas when you are preparing. Even if you only understand *some* of what you have read (in this case, jot down any questions), preparing will ease your brain into the topic, and should help to avoid the panic that can come from

your lecturer rattling off complex concepts at a ridiculously quick pace (a common experience at university). Importantly, this kind of preparation should prevent you getting so discouraged that you 'switch off' in class.

3. Enter with **an open mind and positive attitude** (Schmitz, 2012). Being motivated and non-judgemental are important components of active listening. Your lecturer is likely to have more life experience and may also have a different worldview to you (your classmates may have too). A lecture therefore offers an excellent opportunity in which to exercise your curiosity and learn more about the way other people see the world. You may agree or disagree with what is said during the course of your studies, but try to be open to new ideas, adopt a little appreciation of relativity where necessary, and make notes of any moments of *cognitive dissonance* (visit Section 8.3 for more on this). Try to resolve issues of dissonance after class as you revise the lecture.

4. **Pay attention**. Give your lecturer your undivided attention. Make a conscious decision to 'switch *on*' in class. Easier said than done? Bakker and Niemantsverdriet (2016) say that this is difficult because we have limited mental resources with which to manage the many activities we need to perform. For example, if your lecturer is not engaging, there is noise that makes it difficult to hear, or you are distracted, it is easy to lose focus. If you do feel distracted, try to ignore the siren call of what is pulling you away by refocusing. You can do this by getting rid of the distractions (like your phone, or moving away from chatty classmates), asking yourself a question about the point your lecturer just made, by trying to visualise what she is saying, or by offering some non-verbal feedback. This includes making eye-contact, sitting up straight, smiling briefly, mirroring their expressions, and reflecting your interest (or even confusion). Apart from boosting your comprehension, the bonus of giving your full attention is that it shows you to be an ethical, respectful listener (which may come in handy later, when your lecturer might be more inclined to go the extra mile for you).

In Greek mythology, the **Sirens** were sea nymphs who called unsuspecting sailors onto the rocks or into dangerous waters by singing to them.

5. **Take summary notes** whenever you can – this applies to any important points your lecturer or a classmate makes, and to any presentation content offered. You may find it difficult to identify the key ideas of class topics at first, but you will find summary notes get easier with time and practice. Pay attention to signals from your lecturer here: repeated points, changes in tone or voice, or key words on the board or presentation often indicate important content. Your summary notes can form the basis of your study notes, so make a concerted effort with them by following the advice for summarising in Section 7.4.3. And for clarity, remember to include illustrative examples that helped you to understand what was discussed in class. Good examples with detail will often trigger recall, so can help you in answering test or exam questions later on too.

6. During lectures, **apply a critical lens**. As we noted in Section 2.3.5 on reading, being critical means that you consider what is said (or not said) more deeply – it does not mean that you need to find fault with what your lecturer is saying or that you should be rude (please don't!). For example, being critical can mean thinking about your lecturer's use of tone, persuasive, emotive, or factual language, and noting if there is any evidence of bias in what is said (this is *not* necessarily bad, but should be considered). Schmitz (2012) suggests asking yourself about the content covered in the lecture too – is it non-fiction, factual, based on good theory, or based on assumption? Depending on the purpose of your class, the nature of your content will (and should) differ, but it should always meet your course goals. As a result, you should also consider the extent to which the nature of the content – and the topic discussed in the lecture – is suitable or credible for your subject.

7. **Ask questions** to clarify or expand what has been said **but be patient** – if you want to ask a question, write it down so that you remember it, and then wait for the right moment: you don't want to be so consumed with identifying the time to interrupt politely that you forget to listen to what is being said. It is also important that you ask meaningful questions that respect your lecturer and classmates: for example, asking a question about something off-topic, or about something that is already covered in your course guide, textbook, or by Prof Google is not helpful. Neither is peppering the lecturer with a hundred questions, just for the sake of 'asking questions' (these will have the added 'bonus' of earning you eye-rolls and groans from your classmates). Instead, questions should be thoughtful and show that you have prepared and listened to the lecture, and that you now simply need a little more clarity or depth. Importantly, if your questions meet these criteria, you should never feel too shy to ask them!

8. **Reflect** on your lecture by relating the new ideas to ones you already have – think back to Section 9.2.1 on modelling to see how this can help you to understand and remember new information. As part of this process, revisit your notes and try to mentally sort the information into: a) what is clear, b) what is confusing, c) what is new, d) what you already knew – this metacognitive activity (see Section 1.1.4 for more), which can become automatic in independent learners, should help you to process and recall new ideas, and should also help you to identify any questions you need to ask.

9. Finally, try to **discuss** lecture topics, points of confusion, or possible questions in your class networks or study groups before or after your lectures. You can do this at some comfortable spot on campus, but social media apps like WhatsApp's group function can also be useful. Discussions allow for different perspectives on what is covered, allow you to test your own understanding of what was covered, and should also help to clarify tricky or complex

topics – just make sure that you resolve any issues identified in your discussion with the lecturer or tutor as soon as possible.

13.6 Dealing with different types of questions

Okay, after following the advice in Section 13.4, your walls and bathroom ceiling are covered with yellow sticky notes, your bedroom is full of colourful maps, timelines and spider-diagrams, and your computer is depressingly free of any Facebook, TikTok, Netflix and Reddit – you have been a study machine and are ready to claim your status as an academic superhero. Just one thing … when you were dutifully going over past papers, you realised that there was a huge variety of question types that the exams expected you to be able to deal with. You feel down with **essays and long questions** – you have mastered those through your work on assignments, but what about Multiple Choice Questions (MCQs): those multiple option questions that everyone else seems to love, but which you find annoyingly tricky? Or those awful practical exams in the chemistry lab?

This section will outline some useful strategies on dealing with MCQs and practical lab exams, and help you to work out a plan of action when you encounter them. For all types of questions, remember the principles behind managing your examination (see Section 13.3).

13.6.1 Multiple choice questions (MCQs)

MCQs usually consist of a statement or question, followed by one or more correct answer(s), and three or four distractors (incorrect answers, or correct statements that are not relevant to the given statement). Many students think that MCQs are easier to answer than written response questions, but they can actually be trickier. This is because examiners may mislead students deliberately by using a slightly different wording, or a deliberate misspelling, in the distractors. Students can easily fall into traps if they do not carefully read each question and the options underneath it. The idea with MCQs is to test students' attention to detail and a wide breadth of content. To deal with MCQs effectively, follow this approach:

1. First read and **follow the instructions** carefully – if you need to respond using a particular pencil (often an HB), make sure that you use it; if you are told to complete your answers in a separate booklet, or the back of your exam booklet, do not write them out in your booklet. Regardless of how correct your answers may be, if they are in the wrong place, they will not be marked. Also note whether negative marking will be used – if it will be, avoid taking a guess if you are unsure of the answer. If negative marking will not be used, then provide an answer for *all* questions, even if you have to guess on some of them!

2. Your next step is to **read through all of the questions** to get a general idea of what they entail, and to note the questions that you can easily answer with some certainty. Do not make any marks on your answer sheet yet, but make notes on your question paper if this is allowed.

3. Go through the MCQs again, and **check the answers** that you were sure of in your first run-through. If you are satisfied that you have the correct answer, complete these on your answer sheet, but do not fill in any answers for questions that you are unsure of. Caution: ensure that you don't mix up the answers on the question sheet – the answer you complete must correspond with the actual question you are answering!

4. Your final run-through will include figuring out the more challenging MCQs. Some useful approaches to these are to:

    > use the **process of elimination** to rule out incorrect options (check for obvious jargon errors, use of negative or positive in the wrong position, irrelevant responses, dating issues etc.);

    > if the answers are confusing you, look at the question itself, and try to think about what you would have answered **without the options being present** – then look for the answer that most fits with your understanding;

    > in calculation-type MCQs, try to **do the calculation independently** of the answers and see what answer you come up with on your own – usually, these kinds of MCQs are worth more than 1 or 2 marks because they require more time – just watch that you do not spend too long working these out;

    > ultimately, if you cannot eliminate all of the distractors, you may be required to **guess** – not ideal, but if negative marking is not being used, there is still a better chance of you getting the mark for the question than if you leave it open, especially if you have managed to eliminate some of the options.

Finally, remember that you should consider the time available when you are responding to the MCQ section. Unless otherwise specified, exams usually use a **mark-a-minute principle**. So, if an MCQ is worth one mark, do not spend more than a minute on it, or you risk not getting to your essay questions. The best way to do this is to practise answering MCQs from past exams, checking that you can answer them in the allocated time.

13.6.2 Practical laboratory exams

Usually, the only way that you would have qualified for a practical exam is if you had attended a minimum number of laboratory practicals during the semester. Hopefully, this means that you can find your way around the lab you will do your exam in, and that you are already familiar with the techniques that you will need to demonstrate. For example, in an anatomy exam you might be required to examine a series of histology slides, and then explain, draw and label them. Or you may be required to perform a physical science experiment and carefully record your data and results, explaining your findings to an assessor. Whether you are in a biology, anatomy, geography, physics, chemistry or other kind of lab, the general idea is that you will need to demonstrate competency during a practical experiment or observation, and then present your results to the assessor.

This means that you need to:

- make sure that you are **dressed and equipped appropriately** for the lab – follow the rules provided by your lecturer here, but think about anything you might need, e.g. safety shoes, masks, glasses, calculators, pencils, or protractors;
- work **neatly** and **stick to the rules of the lab**;
- clearly understand, *before you go in*, what **kind of tasks** or experiments you might need to engage in (this should be clear if you have attended your practical sessions, and have studied the necessary theory beforehand). Ask your lecturer for some insights during the semester so that you can get comfortable with the procedures. It is also helpful to make a list of questions that you might be expected to answer beforehand. Practise going through likely activities or questions with your lab partner;
- if you have to respond orally to an assessor's questions, make sure that you **think about the question first**, carefully check what you are looking at, and then take your time when answering;
- if you need to describe a given specimen, **start with the simple**, obvious things (we often overlook these, thinking that they are too obvious). Once you have explained the simpler things, you should feel more confident, and are likely to struggle less explaining more complex things;
- **justify any conclusions** that you have drawn – if you have a good grasp of the theory behind the lab work you have done, this is the time to use it to your advantage. Practical lab exams often award marks for justifying your conclusions, so, for example, if you argue that the muscle filament specimen is eosinophilic, explain *why* you say so (i.e. the pink/purple colour of the specimen was clearly visible);
- for **open book assessments**, ensure that you are familiar with your text beforehand – wasting time in the practical exam looking for things is sometimes worse than not having the text at all, and you could run out of time;
- if you do not know the answer to a question, ask for a hint if you can, or move on to the next topic – **don't fudge** your way through things you do not know. If you babble, assessors will quickly realise that you do not know what you are talking about, and may not give you time on questions you do know the answers to;
- and finally: **be safe in the lab** by knowing your 'stuff' – this is *especially* important if you are doing chemistry practicals or similar. Blowing up the lab *will* be bad for your results.

Before we close with some general advice, we encourage you to remember the processes for writing **essays and long-question** responses that were covered in earlier chapters. These are, by far, the most common type of exam questions that you will be required to answer, since much of university study is about making you think critically about issues. So, please make sure that you revisit the chapters that describe argumentation and writing strategies, and that you are comfortable writing both styles of response.

For all types of exams, remember the following three kernels of advice, *even if you remember nothing else*:

1. When you prepare for your exams, make sure that you can **answer all the subject objectives** that you were given. These may be called different things at different institutions, but they are usually listed under each section of work, or provided in a subject guide. They detail the nature of understanding and the specific areas that you are expected to engage with for the subject (for example, this book includes objectives for each chapter). Your lecturer should set exams based on your module objectives, so they are a great place to focus your studies, especially if you do not have past papers on which to practise.
2. **Read and analyse questions carefully** before you begin, check mark allocations, and answer the questions you have been asked – not the ones you wish you had been asked.
3. **Start with the easy questions** first to build your confidence. It also ensures that you get all marks possible for the work you do know well. Just make sure that you clearly number each answer to ensure the marker knows what they are marking.

Note!

Use the material on the Learning Zone to practise the specific kinds of short questions and long questions that you are likely to get in your faculty. For short questions, see student exercise LZ 13.1, and for long questions, use student exercise LZ 13.2. Your lecturer will be able to give you the answers for these.

13.7 Conclusion

Although exams and studies in general can be daunting, we hope that this book has lessened your anxiety. If you engage with the material throughout the semester, and are able to synthesise the skills that you have learnt throughout the course of this book, you should get through any difficult work or exams successfully, without much trauma.

In this chapter, we highlighted the importance of preparing for your exams by using practical ways to relax, and by ensuring that you manage your time effectively. However, avoiding procrastination is not a one-off lesson: you will probably encounter and deal with this little monster throughout your life. The most important thing is to stop procrastination becoming a habit. Use the tips for guarding against it whenever you feel it creeping in, and you should be able to skirt past it successfully.

The final focus in this chapter was a set of simple strategies for dealing with MCQs and practical laboratory exams, and to remind you how important it is to respond to essay and longer questions by carefully analysing your questions. By considering earlier chapters' work on different kinds of questions, you should feel prepared to deal with a variety of types. We concluded the chapter, and this book, with some practical tips to see you through your exams in one piece, an academic superhero! We hope that you have found this book helpful for your studies. Good luck: may the Force be with you!

References

Bakker, S. and Niemantsverdriet, K. 2016. 'The interaction–attention continuum: considering various levels of human attention in interaction design'. *International journal of design,* 10(2): 1-14.

Caspersz, D. and Stasinska, A. 2015. 'Can we teach effective listening? An exploratory study'. *Journal of university teaching and learning practice,* 12(4). Retrieved 17/09/2019, available at <https://ro.uow.edu.au/jutlp/vol12/iss4/2/>

Cassady, J. C. and Johnson, R. E. 2002. 'Cognitive test anxiety and academic performance'. *Contemporary educational psychology,* 27(2): 270–295.

Chua, C. 2019. 'How to stop procrastinating: 11 practical ways for procrastinators'. *Lifehack,* 16 September 2019. Retrieved 24/01/2020, available at <http://www.lifehack.org/articles/lifehack/11-practical-ways-to-stop-procrastination.html>

Confucius. 1963. *A source book in Chinese philosophy,* translated and compiled by Chan, W-T. Princeton, NJ: Princeton University Press.

Duff, O. 2019. 'Pressure points'. *The Guardian,* 27 March 2003. Retrieved 30/08/2019, available at <http://www.theguardian.com/education/2003/mar/27/studentwork.students>

Hall-Flavin, D. K. 2013. 'Text anxiety: Can it be treated?'. *Harbin Strong,* 12 August 2013. Retrieved 13/04/2020, available at <https://harbinstrong.com/2013/08/12/test-anxiety-can-it-be-treated/>

Intelligent. 2018. 'Create a study plan'. *Intelligent.* Retrieved 30/08/2019, available at <https://www.intelligent.com/create-a-study-plan/>

Martini, F. H. 2005. *Fundamentals of anatomy and physiology.* 7th ed. Upper Saddle, NJ: Pearson.

McGuinness, M. 2020. 'Three big reasons why we procrastinate (and what to do about them)'. *Lateral action.* Retrieved 30/08/2019, available at <http://lateralaction.com/articles/reasons-for-procrastination/>

Psychologist world. 2019. 'Stress test'. *Psychologist world*. Retrieved 31/08/2019, available at <https://www.psychologistworld.com/stress/stress-test>

Sara laughed. 2019. 'Free Library'. *Sara laughed*. Retrieved 24/01/2019, available at: <http://saralaughed.com/free-library/>

Schmitz, A. 2012. 'Stand up, speak out: The practice and ethics of public speaking'. *Saylor academy*. Retrieved 20/09/2019, available at <https://saylordotorg.github.io/text_stand-up-speak-out-the-practice-and-ethics-of-public-speaking/>

ThoughtCo. 2019. 'Mnemonic devices for students'. *ThoughtCo*. Accessed 30/08/2019, available at <https://www.thoughtco.com/mnemonic devices tools-7755>

Weller, C. 2014. 'The photographic memory hoax: Science has never proven it's real, so why do we keep acting like it is?' *Medical daily*, 6 June 2014. Retrieved 30/08/2019, available at <http://www.medicaldaily.com/photographic-memory-hoax-science-has-never-proven-its-real-so-why-do-we-keeping-acting-it-286984>

INDEX